Praise for
The Future of Food Is Female

"A great collection of the inspiring but often untold stories of the women in food tech and how they're shaking things up. Stojkovic's work is a great read for up-and-coming entrepreneurs and innovators interested in the industry."

—Kyle Vogt, CEO of Cruise

"As someone who has dedicated my life to building a better food system, I know firsthand how crucial it is for us to empower the best and brightest minds to solve some of the industry's biggest challenges. Yet, for many of those who are leading the way, their stories have not been told. Stojkovic weaves together a powerful vision for a transformative future of food, as told through the eyes of women around the world. These fellow founders, scientists, and innovators provide a hopeful path for a kind, just, and safe planet for us all, and I recommend this book to anyone looking to be inspired to make a difference in this world."

—Josh Tetrick, CEO of Eat Just

"*The Future of Food Is Female* sheds light on the visionary women who are paving the way toward a more sustainable, healthy, equitable, and compassionate future by challenging the status quo of our food systems. A must read to understand who and what are behind the positive changes taking root across the food industry, this book will inspire readers to see how they can help address some of the most pressing challenges facing our world today."

—Uma Valeti, MD, CEO of UPSIDE Foods

"I am a friend and fan of many of the featured trailblazing women who are dedicated to improving the lives of so many around the

THE
FUTURE
OF FOOD
IS FEMALE

REINVENTING THE FOOD SYSTEM
TO SAVE THE PLANET

JENNIFER STOJKOVIC

VWS PRESS

Published by VWS Press, Los Angeles
veganwomensummit.com

Edited and designed by Girl Friday Productions
www.girlfridayproductions.com

Cover design: Megan Katsanevakis
Project management: Mari Kesselring
Editorial: Bethany Davis

Image credits: cover © iStock Photo/Jolygon; iStock
Photo/Rebelf; iStock Photo/GeorgePeters

ISBN (paperback): 979-8-9857259-0-2
ISBN (e-book): 979-8-9857259-1-9

First edition

For Laika.
The world's first animal to orbit the Earth.
Science failed you, but we will not fail those who come after you.

Contents

knowledge and insights. Where those who are contemplating starting companies can get their questions answered and get inspired. Where investors can meet the people with the next big ideas. And where anyone who wants to can get involved in the future-of-food movement. We are creating accessibility for folks who didn't come from Ivy League schools or from other backgrounds where networking connections tend to come easy. We're providing the platform for the dynamic and incredible women and people of color who aren't being centered elsewhere. And we're writing and placing articles about the people whom most mainstream media are overlooking.

In 2020, when the world entered a pandemic, suddenly everyone cared about the food system. Not only were scientists trying to figure out where COVID-19 came from, but we also had food shortages and supply problems. A massive amount of focus shifted to the industry, creating a record year with more than $3 billion invested in alternative protein. Yet with all that money flooding the market, investments in women founders *dropped*. The reason many investors gave was their perception that it was safer to bet on male-led companies. That kind of ignorance is absolutely unacceptable. With all the challenges currently facing us, we need all hands on deck, including and especially women's hands. We must once and for all get beyond this idea of women as second-class leaders. In fact, this idea couldn't be less true—studies demonstrate that women founders perform 63 percent *better* than all-male founder teams. And the intentions of women founders are decidedly more mission-driven than their male counterparts, with male founders being eight times more likely to start a company for financial gain than women founders.

Investors must do the work of seeking out founders *where they are,* because I guarantee they are out there, and some of them will be responsible for the next revolutionary food-tech breakthroughs—but only if they get the funding they need to develop and scale. Investors must understand that women founders and founders of color often don't have the same connections and other affordances their White male counterparts enjoy, plus many use different platforms for their communication. (For example, lots of investors scour LinkedIn for potential deal flow, yet Black founders are often more likely to be active on Instagram.) If you're one of those investors who says you'd like to support diverse founders but you're unable to find them, I invite you to reach out to me directly, because there are literally thousands of them out there, and I would be more than happy to help you connect with them.

It's not just investors who have blind spots. In the media, people continue to talk about plant-based and cell-based companies, yet they overlook women and people of color who are taking the space by storm. That, in my opinion, is a huge part of the problem—the lack of high-level media coverage for women founders and founders of color. That's why I created this book—to share the stories of these founders and other food-industry trailblazers more broadly.

The Future of Food Is Female represents the first time many of the women in these pages have ever been interviewed for a book. When you read their incredible stories, you'll realize what a disservice that is to the revolutionary work they're doing. One day, this will be a history book whose pages you'll be able to flip or scroll through to remember how it all began—but today, it's telling a story that precious few people have heard. Mark my words, though: for the women in these pages, and for countless others we work with at VWS, this is only the beginning. And in these pages is likely the next Steve Jobs or Elon Musk (only they're working to save the planet we're already on).

In this book, you'll find voices from women around the world from a variety of circumstances, who were brought to this space by very different paths. These women have not only diverse cultural and ethnic backgrounds but also a wide range of skill sets and ways they're impacting the space. Their stories show that you don't just have to be a founder, and you don't just have to come to the food space through a science or tech background—there are so many ways to be part of this revolution. One of the things that most excites me about this book is that young people, regardless of their gender or background, will be able to read these stories and realize, "I can become a leader, too. I can make a difference." There's a path to becoming a leader in this space, no matter where you're from. And we will be here to support you, because as fast as this space is growing, we need more of you. More women, more people of color, more people with your unique insights and skills.

Perhaps the biggest takeaway from this book is that no matter who you are, you can help save the planet, simply by the choices you make about what's on your plate. It's not about making a huge proclamation for yourself or adopting a label—those things really aren't necessary, and it's much simpler than that. In the end, it comes down to a choice we make several times each day. Each time we eat, we have the opportunity to save the world. To ask ourselves: How is what's on my plate affecting the planet? It also includes the clothes we're wearing and the products we're using on our body. When each of us understands, accepts, and takes responsibility for the fact that our lives are composed of daily choices and decisions about what we consume, that's when the real change happens.

I'm so excited that you're taking this journey to the future with us. Now, let's get started.

Chapter 1

Suzy Amis Cameron

Author, Environmental Advocate, and Investor

Reaching Supermodel Stardom, Learning to Love Your Body, and Changing the Way We Eat, One Meal a Day

> "My personal mission statement is to inspire
> as many people in the world as possible to
> go plant-based, to make the world a better
> place for all of our children to grow up in."

Like many, Suzy Amis Cameron and her husband, Hollywood director James Cameron, see plant-based eating as an essential element of our effort to save the planet. After switching to a plant-based diet nearly a decade ago, the couple has become extremely active in the plant-based food sector, founding Verdient Foods, a visionary Canadian company that produces protein from plants for use in consumer food products, and now helping to overhaul aspects of New Zealand's agricultural industry. For Suzy, however, switching to a vegan diet and lifestyle has also netted a surprising benefit: helping to heal her longtime struggle to have a healthy relationship with food.

"I was always around nature," Suzy Amis Cameron says of her childhood in Oklahoma City. She recalls fondly how, growing up, she and her five brothers and sisters spent significant time at their family's farm, located between Oklahoma City and Dallas, Texas. "We were running around barefoot, hanging out and just being Okies," Suzy says.

Small-town life did not last long for Suzy, however, and her path took a dramatic turn when, in her teens, she was presented with the opportunity of a lifetime. At the age of seventeen, Suzy was offered a contract with a major modeling agency. Suddenly, she traded the pastoral scenery of the American Midwest for the bustling streets of Paris. "My whole world and my brain expanded and exploded learning about all of these new things," she says. "I'm so grateful for the experiences that I had and everything I was exposed to traveling across the ocean and around the world." For Suzy, though, her shift to modeling also sparked a newfound confusion around food.

Back in Oklahoma, Suzy had eaten the typical fare—cows and pigs from the family farm, lots of casseroles, and few fresh veggies, though Suzy says she delighted in vegetables and salads when they were on offer: "I remember finding French-cut green beans in the freezer one day and cooking them up, and it was like a revelation for me. We didn't have a vegetable-heavy diet. When I went off to be a model in Paris, all of a sudden you have to think about what you're eating and fitting into the clothes."

Suzy says it was challenging navigating the rich, decadent cuisine of France to figure out what she should eat. "I was this shy and ridiculously naive girl from Oklahoma; I'd never been out of the country at all. Today models get help and coaching on nourishing yourself," Suzy says. While in Paris, though someone from her agency did try to make sure she and the other models were drinking water, eating salads, and getting exercise (including many outings roller-skating to the Eiffel Tower), there was little specific guidance other than that food was largely something to avoid. Suzy recalls an agent making a derogatory comment about her body: "I pulled her aside at one point and said, 'Can you give me some advice on losing weight?' I really didn't need to lose weight, but clearly she thought I did. And the agent looked at me and said, 'Just quit eating.'"

When Suzy made the switch to acting, she found further pressure and few solutions: "It was the same type of thing there, with the focus on my physicality. It was so confusing trying to figure out how to stay in shape and how to be healthy." As Suzy acknowledges and so many of us know, you don't have to be a model or an actor to struggle with food: "I remember growing up, my friend two doors down, her mother always had a can of SlimFast on her desk. There's a dark side to that whole paradigm of eating, and we all deal with that every single day as women, I think. Hopefully we're learning to break that paradigm."

Suzy's shift in her relationship with food started when she was pregnant with her first child and living in Los Angeles. Her sister-in-law took her to a local health-food store and introduced her to organic food. "When you're growing a little being inside you, you start to get very specific about what you're eating," Suzy says. "Then when that baby comes out, they're this vulnerable little thing and you think about everything you're putting in them, on their skin, what you're dressing them in, what's in your room. I think motherhood is the piece that pushed me in the direction of really looking at all of the products I brought into the house."

After being a single mom for six years, Suzy met her husband, director James Cameron. As a family, they held a dedicated diet focused on eating organic fruits and vegetables and consuming only meat from grass-fed cows and free-range chickens. "At the time, I thought it was the best way to eat," says Suzy. In 2005, she cofounded Muse Global School alongside her sister Rebecca, and they fed the students there similarly. Then Suzy's story took what's become a common turn for many who've switched to plant-based eating—in 2012, she and James watched the documentary *Forks Over Knives*, which focuses on the health aspects of plant-based eating and its comparison to the American standard of a diet high in animal products. "We did a one-eighty literally overnight," says Suzy. "We watched it on May 7, and by May 8 our whole kitchen, pantry, everything, was totally cleaned out."

For Suzy, the switch to plant-based eating resolved her confusion over the advice she'd gotten previously, which was to eat a lot of meat and cut out carbs and fat. To her, the idea of an organic, whole-foods, plant-based diet made so much more sense. "For the first time in my life, I learned how to really nourish myself and feel good every single day," says Suzy. After she went plant-based, she was astonished at how much more energy she had and the fact that her weight and how her clothes fit no longer fluctuated from day to day. "I'm in better shape now than I was when I was seventeen years old, and I've had four kids," Suzy laughs, adding that it helps that she's also always enjoyed exercise. "I have an enormous amount of energy and I work out nearly every day, and if I'm not working out, I'm taking a walk or just otherwise staying really active. And the great part is that being plant-based, you can really eat what you want if you're eating healthy. You don't have to overthink it. I feel nourished. I feel full. I feel happy. My brain's alive. And I didn't used to feel that way."

Opting for a plant-based lifestyle also breaks the cycle of Suzy's own and previous generations of women's struggles with eating. "I'm so grateful my kids have a healthy relationship with food," she says. "It's a whole different purview of participating in our world. Going plant-based really changed my life on so many levels."

Suzy's switch started with a focus on her family's health, but quickly it evolved into much more: "It was shortly after that that Jim started educating me on the environmental part of it." She described what many vegans go through, which is the cognitive dissonance of realizing that so much of what you've been told growing up about what you need to eat to be healthy not only isn't good for your health but is also disastrous for the environment. "It was really mind-blowing," she says.

It became apparent that in the interest of their students and the planet, they'd have to change the Muse menu to plant-based options, as well. As a result, Muse became the country's (and possibly the world's) first K–12 vegan school. The school is also zero-waste and completely powered by solar energy. Suzy says that when they announced the changes to the lunch program, many parents voiced objections, but they pushed back on the pushback, underscoring the environmental and health benefits. Plus, it was just one meal a day, and wasn't the planet worth at least that? "When you talk to people about the environment and the massive problems facing us, they can get paralyzed," she says, "but switching your eating habits, even just a little, is an action item that everyone can have."

From there, Suzy and James started looking at how else they might be contributing to climate degradation, including examining their investments and divesting from concerns that weren't good for the environment. "We started looking at how we could turn our investments into something that's beneficial for the environment, for humanity, for the animals, and so on," she says. At that point, they were owners of one of the most successful dairies in New Zealand, which they shuttered. As an alternative, they started buying farmland in Saskatchewan and focusing on protein crops that grew well there, such as yellow peas, lentils, chickpeas, and fava beans. But they didn't just want to be farmers; they wanted to think further into the future and down the supply chain: "We wanted to add value to the process, so we started a relationship with the Food Centre at the University of Saskatchewan to create food products using the plant proteins we were producing." The result was Verdient Foods, which manufactures consumer packaged goods (CPG) such as pastas, sauces, and butters. At the time, consumer focus on plant-based foods, especially as it pertains to investment in the supply chain, was still very early. Suzy says people were a bit puzzled about the idea of growing plants for protein, "but they get it now."

The Camerons have since sold Verdient and are now focusing on supporting the plant-based agricultural revolution in New Zealand, where the family has taken up residence while James focuses on his next filmmaking projects. As part of this effort, they've built a massive farming concern focused on growing brassicas, a plant genus that includes broccoli, cauliflower, cabbage, kale, and brussels sprouts. "We're now the number one organic-brassica grower in NZ," Suzy says proudly. As with their efforts in Canada, the pair want to do more than produce commodity crops. They're focused on rehabilitating the land, as well as providing New Zealand's dairy farmers with profitable plant-based alternatives.

"New Zealand is so dairy-intensive, so we're focused on helping farmers find a way to pivot away from dairy and have an alternative that is just as lucrative, if not more lucrative than, dairy. In the process of producing food, we want to be able to build soil instead of degrading it, and clean the water instead of polluting it," she says. "New Zealand is so forward-thinking in so many ways, and they know they can't meet their Paris Agreement levels without addressing animal agriculture." To this end, the country announced that they would reduce animals in their agricultural system ("livestock units") by 15 percent by 2030. "It's not enough," says Suzy, "but it's something, and I don't know another country in the world that's talking about reducing livestock units. And if Jim and I can have any influence at all in helping farmers find something else to grow or do that is just as lucrative and that adds to the GDP in New Zealand, that's what we're focused on."

Suzy says dairy doesn't always add up financially for farmers, and can even leave them worse off: "What's really devastating is that these dairy farmers get in debt over equipment, more inputs on their grass, and so on. They're also having to inflict this horrible handling on the animals, and that takes a toll." Suzy points to the surging suicide rate among New Zealand farmers, a trend that is mirrored in several other countries.

As part of this effort, Suzy became an executive producer for *Milked*, a documentary about the dairy industry in New Zealand for which she was also interviewed. "It's walking a tightrope because it's looking at the deep darkness of the dairy industry," says Suzy. "I knew it was really bad, but I don't think I really knew just how horrible it truly is, like what happens to the baby calves who are culled. It's just devastating what the dairy industry does to the babies and the moms. You drive down the road, and you see on one side of the road these little babies, and on the other side you see all the moms full of milk, and those babies are not getting that milk. A typical female cow would live about twenty-five years, yet in the dairy industry their lifespan is only about four to five years because they're kept perpetually pregnant and it just takes such a toll on their bodies," Suzy says.

It's also about the harm the industry is doing to our environmental and personal health, which are inextricably linked. "We're living in climate change," Suzy says. "In New Zealand, within a two-week period we had a thirty-year storm and, a week later, we had a fifty-year storm. There are floods, droughts, and hurricanes. New Zealand is also number one in the world for colon cancer, and everyone says it's the nitrates in the water," Suzy explains. "What people don't necessarily put together is that those nitrates are there because the farmers are growing grass to feed the cows, and those soil inputs leach off into the water. But it's not just the nitrates; it's also the diet people are eating, which is rich in dairy, which we know turbocharges carcinogens like aflatoxin."

Their efforts are paying off. In addition to the Camerons' block-buster brassica operation, Suzy points to another former dairy farmer who has met with success switching to growing pumpkins and another to hemp.

Yet for Suzy, that's still not enough. "My personal mission statement is to inspire as many people in the world as possible to go plant-based, to make the world a better place for all of our children to grow up in," Suzy says. "I go to sleep with that, I wake up with that, I'm up in the middle of the night with that. I'm thinking, 'How else can I use my gifts and talents to make the world a better place for all of our children to grow up in?' The grownups are the ones that have made the mess. When we're long gone, our kids are going to look around and say, 'My God, why didn't they do anything to make a difference? Why didn't they make changes?'"

As Suzy's quick to add, though, things are looking up, with even large corporations now starting to look at their supply chains and the impact they're making. Personally, she has joined the board of directors of the Livekindly Collective, a collective of plant-based entrepreneurs and brands working toward a sustainable food system, among others. She's also authored the popular book *The OMD Plan* to inspire people to switch their diet to include at least one plant-based meal per day, which has led to conversations with high-profile leaders and philanthropists around the world, like Oprah.

Suzy predicts that a major change in how we view food is just around the corner. "When I think about animal agriculture," she says, "it's like the next cigarette. There was a time when people smoked freely. I remember a smoking section in the back of the plane when I was flying back and forth to Paris. That was the norm. Then little by little we started to realize that secondhand smoke was a problem, and smoking wasn't hurting just smokers but the rest of us, as well." Similarly, she says, someone's personal choice to eat a burger or steak is causing harm to the environment, and that's affecting us all, including our health.

"It's a fundamental shift we have to make with food," Suzy says. "We need to think not only about ourselves but also about what's best for us as a species to survive. Will the change happen in my lifetime? I sure as heck hope so. I think things happen faster now than in the past because of the type of connectivity we have."

Suzy says one of the things that excites her most is how many women have been inspired to get into the plant-based movement. Being more willing to work together toward a greater goal can help us get there faster, she explains. "Most women's brains are wired to cooperate instead of compete, so it's important to have women in this game to move us forward."

When I ask her advice for women and other entrepreneurs in the sector, she hearkens back to an approach they teach at the Muse school. "We call it ORP, and it's that you stay *open*, you stay *resourceful*, and you stay *persistent*. You need to be able to hold steady and keep moving when challenges arise." Suzy says that all too often, women give up too quickly: "When someone says, 'I went up to the front door and they said no,' I'm like . . . is there a window? Is there a back door? Is there a cellar entrance? You just find a way." Fortunately, Suzy says, "I think women are hearing yes more often than in the past. We just have to remember that there are ways to make things happen."

Chapter 2

Pinky Cole

CEO of Slutty Vegan

Overcoming Failure, Building a Fast-Food Empire, and Leading the Black Plant-Based Revolution

"I like to help people with no reservations.
I want to see people win."

The path of Pinky Cole is an unbelievable, incredible, and inspiring tale, beginning from day one: on the day Pinky was born, her father was sentenced to more than two decades in prison. A lifelong entrepreneur, Pinky has amassed one of the fastest-growing plant-based empires in the world with Slutty Vegan, a restaurant and CPG brand focused on serving underserved, predominantly Black and Brown communities. Based in Atlanta, Georgia, Pinky credits her meteoric rise to one simple concept: community.

Like so many incredible women entrepreneurs, Aisha "Pinky" Cole comes from modest beginnings. Growing up in Baltimore, Maryland, Pinky was raised by her mother, along with her two sisters and two brothers. On the day she was born, Pinky's father was sentenced to prison for what would turn out to be more than twenty years. "I'm a mix of both of my parents," Pinky says. "I would watch my mom work every single day. My dad was a hustler, and my mom was, too, but in a different way. She worked at a bank, at McDonald's, she did whatever she could to make ends meet. What my dad did wasn't totally legal, but he was a very smart businessman."

Both of Pinky's parents immigrated to the US from Jamaica. Her mother came to the country as a dancer on tour, but when she met Pinky's father, she decided to leave her family in Jamaica and establish roots in America. When she was growing up, Pinky's household was bustling, not just with her and her siblings but also with the other Jamaican immigrants to whom her mother provided shelter and help to get settled. "Now that I'm an adult," Pinky says, "I realize that's why I want to help others—I got it from my mother." In spite of the many demands of meeting the needs of the kids and providing a home for others, Pinky's mother always managed to put food on the table. "We didn't always have everything we wanted," she says, "but we had everything we needed. We didn't have much at Christmas, and I didn't really grow up with toys, but we had the necessities."

In the midst of all of this, Pinky's mother was also a singer in a reggae band, and Pinky recalls going with her to shows growing up, hanging out at a hotel or in the venue's basement while she performed: "Everybody knew my mother—she was a local celebrity. She was never afraid or shy to talk to people, and I learned that from her." Pinky also followed her mother's way of eating. "I grew up differently from the typical Black American. Because we're Jamaican, I didn't grow up eating cheeseburgers, which is funny because now I own a cheeseburger restaurant. My mom's been a vegetarian all her life, so I ate what she ate. I grew up eating home-cooked meals for nearly every single meal."

Like many women in the entrepreneurship world, Pinky also had another influential woman in her household growing up: her grandmother. "My grandmother was very regal and sophisticated, quiet and observant," Pinky says. "My mother was loud—you were going to hear her before she walked into the room, and she doesn't hold anything back. I got my grandmother's quietness, but I got my mother's attitude and confidence, so it allows me to be humble and confident at the same time. My mother taught me how to cook, to work hard, to be consistent, and to do what I loved." Pinky says she also got something else important from her mother—financial literacy. She describes her mother as always being on top of their money situation, understanding debt, and so on, which Pinky says was extremely helpful for her to see because she wasn't given any kind of financial education in school.

Pinky's early years were further enriched by spending summers in Jamaica, where she was able to get in touch with her own family's roots: "We are a village-driven family. All of my aunts would help take care of the kids. Before my grandfather passed, he was the patriarch of the family and my grandmother was the matriarch, and everyone grew up in a big fourteen-bedroom house. When you'd come of age, you'd have your own children and raise them in the house. Those humble beginnings really taught me about life and still help me maneuver through life today." Pinky also says that her time on the island helped shape her attitude toward resource consumption. "Growing up in America, you have access to more resources than you have there," she says. "In the summertime, you'd take a bath outside and would have to go and get the water with buckets," she recalls. "How I clean, how I cook, how I try not to waste, how I think of things—all of that was influenced by my time in Jamaica."

Pinky's foray into the business world started early, in high school. Unlike many of us, when recalling her teenage years, she looks back fondly upon the time—thanks, no doubt, to her fun-loving nature and entrepreneurial spirit. "High school was really good to me," she says. "People would hear the name Pinky and think, 'Who's that girl?' and they'd want to get to know me. I'd go to a lot of parties and work as a promoter making thousands of dollars a week before it was really a thing." Pinky says she was always a bit of an old soul in that even as a kid, while her peers were preoccupied with typical pursuits, she was always thinking about how to build something and get ahead in life. In these early years, her entrepreneurial older sister was also an influence, paying her own way through school and buying her first car using money she earned doing hair. "Growing up," Pinky says, "I really jumped on the entrepreneurial train. That's where I always felt most comfortable."

High school wasn't without its bumps in the road, however. Pinky was expelled from her regular high school for a dispute with a classmate over the voting for prom queen (a title Pinky says she won), but, being a problem-solver, she worked it out: "I was out of school for a month and wrote a letter to the superintendent. They allowed me to finish my schooling at an all-girls school. It was definitely culture shock, but I ended up meeting one of my best friends there." Pinky says another unexpected benefit was that her experience at her new high school inspired her to want to attend Clark Atlanta University, the first historically Black college and university in the Southern US: "It was the best decision I ever made in my life, because I was able to connect with so many people that I never thought I would connect with. I became part of a sorority, the queen of the school, and I just built really good relationships with some of the people who support me now in my business."

Pinky didn't seek to become a business owner immediately, though. Initially, she aspired to become an actress, so she moved to Los Angeles with a duffel bag, a suitcase, and $250 to her name. While the acting part didn't work out, one of Pinky's sorority sisters introduced her to a TV producer who needed help. As I've seen over and over again, one of the things that often distinguishes successful entrepreneurs is the quality and depth of their relationships. While it's true that some businesspeople try to build and use relationships primarily for personal gain, the most successful entrepreneurs engage and invest in authentic, longer-term relationships—many of which end up paying surprise dividends.

As a result of this TV connection, Pinky ended up producing a number of top-rated shows, and her work in production eventually took her to New York City, where she worked as a producer on *Maury*, an experience she says was amazing. "It was fun because I got to meet people where they were," Pinky says. "This was also a place where I realized that it didn't matter what color you were, that everybody goes through the same things, and that color doesn't tell you what life is going to look like for you. I used that as an opportunity to learn how to encourage and empower people, and that felt really good to me. That's how I knew I was going to be in ministry and help people," she says.

It was also in New York that Pinky opened her first food business, Pinky's Jamaican and American Restaurant, which, although Pinky was vegetarian, featured meat on the menu. "I didn't know anything about opening restaurants—I just knew I wanted to have one," says Pinky, "and I knew I wanted to create an experience that people really wanted to be a part of, and people really supported it." After just two years, however, Pinky was forced to shutter the restaurant when, after a grease fire, she lacked sufficient insurance to rebuild. "I was really sad and frustrated. Life just didn't look like how I wanted it to be," says Pinky, who, after closing her restaurant, was evicted, then had her car repossessed. Still, Pinky says she doesn't regret what happened for a moment. She believes the restaurant caught on fire because it wasn't in alignment with who she was and with her personal values. "I didn't even eat the chicken, and I was selling it," she says. "So, everything happens for a reason, and I'm happy that it happened the way it did because it really made me the woman I am today. Now I'm fully in alignment," she says. "I'm walking my purpose. I'm vegan, I get to serve people vegan food, and I get to teach people about veganism who've never heard about it, and I get to do it in the funnest way possible."

When I ask Pinky how she picked herself back up after the devastating loss of her restaurant, she says, "Up to that moment, everything in my entire life that I'd put my hands on had turned to gold." She continued, "So this one thing seemed like a real failure, and I wasn't sure at first how I'd come back from it. I know it sounds clichéd, but pain is temporary, and problems don't last forever. It took me a while to realize that, because that was probably the lowest I've ever been."

That's where Pinky's fortitude kicked in—she told herself over and over that things wouldn't stay bad, and that the universe doesn't make mistakes: "I knew that this was teaching me something, so I had to pause and reflect on what was there for me to learn. And what I learned was that you go through things so that you can be a testimony for other people. These experiences belong to you, but they belong to you so that you can share them with the world."

Now, Pinky says, there's nothing she can't do: "I can literally do anything I want to do because I went through such a tough experience. When speed bumps come my way I think, 'I've been through worse than this.' Everything else is light work by comparison, so these experiences really do shape us. Plus, going through something like that on a smaller level prepared me to take on something bigger, and that something bigger was Slutty Vegan."

In 2014, Pinky shifted to veganism as part of her desire to be more conscious. "I'm always interested in ways to make myself better and deposit new things into my mind," she says. "Going vegan just felt like the next step toward my ultimate goal, which is to be raw vegan, so I did it, and it was the best decision I could have made." Pinky says going vegan made her more in tune with the world around her, including making her more compassionate to others, including animals. Also, she's proud of the opportunity it gave her to inform and inspire: "To be a Black girl like me in a space where veganism was not celebrated at the time, and being an anchor for it in my own world, let me set the tone among my family and friends for what veganism looked like."

Though lots of people were, at first, confused by Pinky's decision to go vegan, before long, they started asking her advice on how they could go vegan, too. "Community is important to me," Pinky says. "I saw that in my mother. Everyone who needed assistance or support, she was there to help them with no reservations, and I'm the same way. I like to help people with no reservations. I want to see people win." That sense of community support is why Pinky has sought to locate Slutty Vegan franchises in areas that are food insecure or that otherwise aren't as attractive to large developers, starting with her first location in Atlanta. Pinky says it's also important that, when businesses do come into these areas that are majority Black and people of color, the business owners look like the residents.

Pinky's goal is to make vegan food, even if it's vegan comfort food, as accessible to people as regular fast food. "I want people to have access to vegan options," she explains. For one, it's important to her that vegans not feel like they have to suppress who they are in order to have a rich and beautiful experience of life. "As a vegan person, we get tired of going to a regular chain and getting a side salad and some fries. We want real food, just like everybody else, and I think that Slutty Vegan and my restaurant, Bar Vegan, do that." Having accessible vegan options also makes it much easier for people to make healthier choices. "Then, all of a sudden," says Pinky, "people have taken the first step. It's all about doing it in a way that's fun and engaging. When you come to Slutty Vegan, we laugh with you, we play hip-hop music, we'll hug you, and we'll give you the experience of a lifetime so that when you walk away you say, 'You know what, I can try this lifestyle,' or 'I can try adding more vegan options.'"

When people feel relaxed and engaged, Pinky says, they're more likely to be receptive and actually seek out more information about what it means to be vegan. She actually leans on some of her earlier career experience as a producer, where she learned how people tick and how to engage them. "That's why I named it Slutty Vegan," she explains. "If I named it Pinky Vegan, people wouldn't be asking questions about it. Plus, we know that sex sells. It works for everything else, so why can't I sell sex for good? And it's been working."

I ask Pinky if she's ever gotten pushback from investors or consumers. "No," she says, "I really haven't. Plus, I've done so much for and in the community, people really support me. People know I care about them and really give back, so they're not going to give me a hard time about the name, which is really just designed to drive people in." In a lot of ways, Pinky says, the name also helps to counteract the perception that veganism is just for rich White people. "And it's not," she says. "Veganism is for everybody. It don't have a face, and it don't have a color. We bring people together in the name of food. And if we can bring people together, if we can drive that and spark a bigger conversation, then we've won."

Pinky's high-touch approach differs drastically from the direction many restaurants are going in, which is toward ghost kitchens, locations that forego dining rooms and focus on delivery and sometimes takeout or drive-through options. That's actually how Slutty Vegan began: Pinky says she started in a prep kitchen, but it got so popular she shifted to a food truck, then on to brick-and-mortar locations. Ghost kitchens are a great way to get your business off the ground and to succeed because you've got such low overhead, Pinky says. She started Slutty Vegan with just $5,000 and encouraged people to order through a social media app. The venture was an overnight success, almost literally, earning $4 million in gross revenue in its first year, operating out of a 650-square-foot space. Yet when it comes to actually disrupting a market and an industry and introducing people to a lifestyle, the reason she and Slutty Vegan and her other brands are doing so well is, in large part, their high degree of visibility, which is fueled by that community-oriented, personal experience.

When I ask Pinky if her vision is to be on every corner in ten years, like a lot of fast-food chains, she says no, but that Slutty Vegan will be in every food desert. After just a year and a half in business, Pinky did her first fundraising round, netting enough money to open thirteen new locations over the next two and a half years. "Slutty Vegan will be a household name, yes," she says, "but the intention is to deposit this brand into areas where you do not see vegan food, and where you don't see a healthier option or alternative to what already exists. That's all I want to do, and if I can do that, then everything I worked for and put together from day one will be fulfilled, and I will be a happy camper." Pinky maintains that mission is more important to her than money, which she says she could make in any number of other ways if she so chose. "I want to put these restaurants in locations where people have probably never heard of vegan food before, and there are so many people that applies to. People really don't know what's vegan and what's not, not because they want to be ignorant but because they haven't been educated. It's my mission to make sure that people know."

Pinky's not stopping there. When I spoke to her, she was in the process of working on a program to put vegan vending machines into local public schools, with healthier snack and drink options for kids. "I think that's going to be a game changer because it will show people that you don't have to eat junk food all day long. That there are healthier options that are vegan and that taste good," Pinky says. "If we deposit this message into the minds of young people, they'll grow up consciously thinking about the food they eat."

As I know from my background in politics, working with the public school system is, for many, a stepping-stone to a political career. I ask Pinky if she has any aspirations along these lines. She confesses to considering a run for city council, but has decided to focus on food for now: "Really, I'm already a politician. I already do what politicians do, I'm just not in office, and I don't have the title. I can do what I do and help the community out, and it's still a win and everybody's happy." Given Pinky's passion and know-how, I have a personal suspicion that a political career *will* be happening, and sooner rather than later.

In the meantime, though, Pinky's got plenty on her plate, including working on a book for Simon & Schuster aimed at getting people who aren't already vegan interested in a plant-based lifestyle. "It's that cheeky table book for people who are foodies, whether they're vegan or not," she says. As if that weren't enough, she's also working on a vegan handbag and shoe collection with megadesigner Steve Madden, and she's been involved in a documentary chronicling Slutty Vegan's meteoric rise to success and how they've been able to change the conversation around what being plant-based means, including a focus on the growing vegan movement among Black and Brown people. Oh yeah, and she's opening about twenty new Slutty Vegan locations in 2022, on top of being a mom *and* planning a vegan baby-food line.

When it comes to Pinky's advice to budding entrepreneurs, she says you need two key things: a good accountant and a good attorney. "My accountant and my attorney are my best friends," she laughs. "Their number one job is to make sure that my business is protected." Even if you don't need their services every day, make sure you have them available on retainer. "It's one thing to have a brand that's popular and that people are talking about; it's another thing when your books are clean to match that," she says. "If you aren't organized, don't have business insurance, and don't make sure your employees and taxes are paid, you don't have a business." Get your core functions organized and on point from the start, she advises, and you won't have to worry about it later.

Pinky also notes that whatever your product or service, you've got to create something that people want to be a part of. Beyond that, it's essential to believe in yourself and your vision, regardless of what others may think of it: "If you believe in yourself, and you believe in what you want to do, if you want to save the world, you can do it."

Chapter 3

Lisa Dyson, PhD

CEO of Air Protein

A Scientist's Journey to Entrepreneurship, a Love of Math, and Use of NASA Technology to Make Better Protein

"We're excited about the impact we can have in this
sector. There's a lot to do. There's a lot of different
types of meat, and people eat it all over the world, so
it's a big enough problem that we have our hands full."

*It's not magic, but it sure seems like it. How else could a company make
a healthy, complete protein out of thin air? Dr. Lisa Dyson's Air Protein
is a moon-shot endeavor, not only figuratively but also literally. Lisa
was inspired by long-forgotten NASA technology that her company,
Air Protein, is now using to create healthy, sustainable alternatives to
animal meat. But for Lisa, a former Fulbright Scholar with a PhD in
physics from MIT, the inspiration for her space-age goal came out of a
deeply traumatic experience: working on the ground in the aftermath of
Hurricane Katrina.*

A California kid, Dr. Lisa Dyson grew up primarily in the Los Angeles
area, yet she says she spent a good deal of time in Louisiana visiting her mother's family. As a young girl, Lisa was passionate about

mathematics. "I loved problem-solving," she says, "and later when I visited a physics lab researching alternative energy and plasma physics, it was like seeing math come alive."

When I ask Lisa how it was for her being a young Black woman interested in science, technology, engineering, and mathematics (STEM), an area typically associated with boys and men, she says, "Early on there was no real distinction between boys and girls, but as I got more and more advanced the numbers of women decreased. Then as I got into physics the numbers were even smaller." Fortunately, she had a role model to look to within her own family, her cousin: "She's a rocket engineer, so she obviously loved mathematics and physics, as well. Seeing her in high school and in college bringing back her engineering coursework and her drafts definitely made her a role model for me."

When it came to learning business leadership, Lisa looked to her father, who she says at one point was the head of a chain of more than fifty hair salons and was very involved in the fashion industry. "He was very much an entrepreneur, and I grew up with entrepreneurs in my family," Lisa says. "I basically have it in my blood. My dad in particular was very creative. I got to see his ideas come to fruition, and see him unite and motivate and inspire people to complete these big tasks and accomplish these missions that he set out." But, as is typical in business, it wasn't always sunshine and roses. "I saw the highs and I saw the lows," she says. "I saw ideas come and fly, and I saw others not fly. It was all of the ups and downs and ins and outs of being an entrepreneur."

But how did a young girl interested in math and physics make the leap to food science? Lisa says her interest in food started when her focus shifted to climate change. After Hurricane Katrina wreaked havoc on the US Gulf Coast in August of 2005, Lisa traveled to Louisiana to be part of rebuilding efforts. What she saw there left a lasting impression on her. In later years, Lisa says, "I kept thinking back to that devastation that happened because of this significant event, where many people lost their loved ones and lost their homes." Katrina remains one of the most severe Category 5 hurricanes ever recorded in US history, with over 1,800 fatalities and $125 billion in damage, primarily affecting New Orleans. To this day, many residents of New Orleans, which is among the nation's largest Black-majority cities, continue to be affected by the traumatic psychological and economically devastating effects of the storm.

Shocked by the magnitude of Hurricane Katrina, Lisa recalls, "There were climate scientists who kept telling us that these weather events were going to be more frequent and last longer. That's what got me interested in trying to be part of the solution." For Lisa, how to focus her energy was a simple calculation. "The food sector is a larger greenhouse-gas emitter than the entire transportation system," Lisa says, "so ultimately, if you want to have an impact on climate change, we have to think about how we eat."

As Lisa explains, the negative environmental outputs from a cow are equivalent to a car, and when you also look at the fact that just a steak takes two years to cultivate, "it's a hugely inefficient process and creates greenhouse gases on the way. And more and more people are becoming aware of that." Additionally, Lisa points out that traditional meat production contributes significantly to deforestation: "In 2019, Brazil saw record fires, and that was because of deforestation to make room for cattle grazing. So, there are issues that the meat industry is causing, and as we go to ten billion people by 2050, how could we scale the way we've been producing meat?"

Lisa's focus on the connection between what we eat and the climate crisis is an increasingly common viewpoint, particularly among younger generations like Gen Z and even the rising Gen Alpha (the current generation of children born from 2010 onward). Gen Z in particular has a distinct affinity for "eating for climate," largely because they are the first generation to receive comprehensive education around climate change through their schooling. In fact, a recent study found that media coverage confirming the existence of climate change increased from less than 50 percent in 2004 (when most millennials were school-aged) to over 90 percent by 2019. Gen Z is also the first generation to grow up with social media, which has been proven to be an effective tool to increase climate-change awareness (and even public-policy change). Unsurprisingly, a recent study conducted showed that more than one-third of Gen Z consumers plan to stop eating meat entirely, while nearly 70 percent reported being anxious about their future due to climate change. This indicates that while Lisa's focus on targeting diet, and meat in particular, for environmental impact may have been a bit ahead of its time back in 2005, it is certainly a mainstream point of discussion and impact today.

I ask Lisa what her experience shifting her own diet to one that's more plant-based was like. "My mother's from the South, so it was hard, when I became a vegetarian, going to our family reunions and finding dishes that didn't have some type of meat in them. But that was years ago. Last year, I found out that my aunt and uncle, who were both raised in Louisiana, have become vegetarians. I nearly fell out of my chair," she says, laughing. "Some of my cousins have been contemplating switching over, as well." Lisa acknowledges that this is part of a larger, multigenerational trend in society. "People are shifting to more meat-free or plant-based diets usually for either health or environmental reasons. It's the awareness of climate change and the fact that meat is the biggest culprit."

When it came to deciphering how she could address these issues, Lisa had her knowledge in math and physics to lean on, along with a stint as a business consultant at the Boston Consulting Group. While there, Lisa gained experience helping large corporations solve big business problems. "I had those two things in my background, so as I thought about these challenges, and I thought about how I could play a role, I zeroed in on the space of technology in industry," Lisa says. "I thought that if we could come upon a solution that's massively sustainable, and we could make it accessible and available, and consumers liked it, then industry could scale it. It's just one way of scaling solutions to some of the negative impacts of some of the systems we've built."

In 2008, Lisa and cofounder Dr. John Reed created Kiverdi, a research-and-development company that explores how carbon transformation can impact the world. Air Protein operates as a subsidiary of Kiverdi.

"It was a winding road to get here," Lisa says of the past ten years. "What I'm excited by now is the transformation that I've seen over that decade in the investor world and in corporations focusing on sustainability. There's more of an awareness of the need for sustainable solutions and innovation to get us to where we need to be for our planet and our progeny, and there's also an awareness that these businesses can be valuable and profitable."

Among Air Protein's supporters are heavy hitters GV (formerly Google Ventures), Barclays, and Archer-Daniels-Midland (ADM). As the world's largest fermentation company and North America's largest protein producer, it made sense that ADM was interested in Air Protein. As to why Barclays, a company that traditionally hasn't demonstrated a significant amount of interest in the plant-based food space, wanted to get involved, they had just developed a fund that was focused on sustainability, and, Lisa explains, "We were one of the key companies that met their criteria." Additionally, GV was starting to dip their toe into the plant-based arena and saw the idea's potential. "Each of these companies was interested in Air Protein because of its unique approach," says Lisa. "As far as I know, we're the only air-based meat company out there."

At a time when interest in alternative proteins abounds, Lisa says, "I think investors saw our approach and thought it was a good value proposition." The company raised $32 million in a 2020 funding round.

Air Protein's unique technology is based on work done by NASA roughly sixty years ago as part of the space program. The idea for the tech came from looking at how NASA answered the challenge of producing food for the confined environment of a spaceship. "They were thinking about how you make food that delivers the carbon we need to survive. It requires making closed-loop carbon cycles on that spaceship. So, they thought about astronauts breathing out carbon dioxide—capturing that carbon and feeding it cultures to produce protein that the astronauts could eat. Then they breathe out more carbon dioxide, which can be captured," Lisa explains.

As Lisa describes Air Protein's process, "It's very similar to fermentation [such as how beer, wine, or kombucha are made], but fermentation reimagined, so we can make a complete protein in the most sustainable way in terms of its carbon, land, and water footprint." She explains, "Typically with fermentation, when you make yogurt or beer, the culture produces carbon dioxide. Our process works in reverse. Instead of producing carbon dioxide, we use elements of the air to create a nutritious, densely packed protein ingredient." The process combines carbon dioxide, oxygen, and nitrogen with water and minerals and, using a probiotic process, converts it into a protein. The protein not only offers a full assortment of essential amino acids but also includes bioavailable vitamins and minerals. "By applying a culinary technique to the protein, we end up with meat," Lisa explains. Currently, the company produces a protein-rich flour that is converted from gas in its specialized fermentation vessels.

"Nature has these closed-loop carbon cycles, but unfortunately we've been getting too much carbon out of the ground and putting it in the atmosphere faster than nature can absorb it," Lisa says. In many ways, Air Protein is helping to rebalance nature's own systems: "The nice thing about what we're doing is that from cradle to plate, when you get your piece of air steak, it's actually carbon negative. It's the only process out there that's creating carbon-negative meat." The company made the world's first "air-based chicken" in 2019 and since then has expanded into other forms of meat, as well as seafood.

When I ask Lisa what she sees for the next five years for her company, she says, "Only excitement! Now that we've demonstrated that we can use the protein to make a variety of meat forms, we're focused on scaling the technology. That's really the focus now. We look forward to Air Protein eventually being the meat product of choice all over the world. We think people will love it for its taste, for its sustainability and what it's doing for the planet, and for its nutrition."

As Lisa said in an interview with *Food Technology* magazine, "Ten years from now, there will definitely still be people who eat meat, probably from animals. But there are many of us who are scaling up those alternatives so that there are choices out there. And as we see the choices, we'll be making the core ingredients in other ways if the way we define meat is not based on the fact that it's from an animal, but based on the textures and the flavors that you're looking for when you bite into a juicy steak or chicken breast."

Lisa's viewpoint is an interesting sticking point in the world of alternative protein, as regulatory and industry battles are waged over the use of the word "meat" in products that do not derive from a slaughtered animal. Similar to the cell-cultivated proteins that Dr. Sandhya Sriram is working on at Shiok Meats or even the plant-based meats that Shama Sukul Lee is producing at Sunfed, Lisa's Air Protein will likely face harsh scrutiny over use of the "meat" label in coming years. Many industry leaders are already fighting fiercely to redefine these terms to better reflect the incoming reality of food technologies and advancements such as Air Protein.

Lisa's vision remains remarkably determined and optimistic—how can it not when you are literally resurrecting space-age technologies to solve humanity's greatest crisis? "We're excited about the impact we can have in this sector," Lisa says. "There's a lot to do. There's a lot of different types of meat, and people eat it all over the world, so it's a big enough problem that we have our hands full."

In 2019, Lisa was named to *Inc.'s* Female Founders 100. When I ask her advice for other women thinking about starting their own companies, she is enthusiastic. "If you want to do it, just do it," she encourages. "It's not going to be a linear path; it's going to be a windy road. There are going to be some bumps along the way. Find your north star—where your head is—and that will help brighten your path along the way." She adds, "One of the things I'm continually amazed by is how many people want to help. Find your advisor, your group of people who have knowledge you don't have, and your people who will just be your cheerleaders."

Chapter 4

Michelle Egger

CEO of Biomilq

Promoting Socially Impactful Capitalism, Being a Millennial CEO, and Empowering Mothers Through the Future of Infant Nutrition

"It's becoming more and more normal to bring your passion and beliefs and cares to the work you do and to try to find impact in everything you work on."

For Biomilq CEO Michelle Egger, building a cell-based breast-milk company isn't just about creating healthier, more sustainable options for nurturing the next generations. She forecasts that companies like hers will help to flatten the structure of the global food system, redistributing power in ways that could enable women in even the smallest, most remote villages to become powerful actors in their own future. By creating choice for mothers around the world, Michelle sees her role as more fundamental than producing healthy, nutritious products: she views it as empowering women.

Unlike many women trailblazing a path in STEM, being a scientist is in Michelle Egger's blood. Her mother was an environmental chemist who opted to stay home to raise Michelle and her sister, who themselves both went on to follow in her academic footsteps. Never shy of

overachievements in school, the two forged their respective paths—
Michelle's sister became a PhD plant biologist, while Michelle became a food scientist. Michelle recalls that as kids, when they'd ask
typical childhood questions such as "Why is the sky blue?" instead
of giving the response most of us would receive, their mother would
give a detailed explanation about light refraction and the density of
air. "As women we were empowered to think about scientific careers,"
Michelle says. "We felt really supported and stimulated in a lot of in-
tellectual ways."

Entrepreneurship came naturally to Michelle, as well, with
Michelle's father, who ran the family's office-furniture business, at the
other end of things. His theory was that it didn't really matter what
the answer was, and that what was important was whether you could
convince someone else of what the answer was. As a young girl, she
remembers watching him practice his sales pitch as he'd get ready in
the morning: "As a teenager I was like, 'That's so weird, Dad!' and now
I get it." Michelle also recalls a moment when she was having a dis-
agreement with her sister and he pulled her aside to say, "Everyone
is your customer, Michelle. If you treat everybody like a customer to
some extent, you'll build better relationships, and you'll have more of
a foundational experience together." Michelle pushed back at the time,
but now she sees the wisdom of his advice—frequently she asks herself
how she can build relationships in ways that help others and that make
them feel comfortable having open conversations.

She says today, both of her parents can empathize with her struggles: "My dad understands what it's like to wrangle with lawyers, and
my mom asks me questions like whether we're able to get the lab supplies we need given the restrictions of COVID-19. Not everyone gets to
have those kinds of conversations with their parents."

Growing up in Minneapolis, Minnesota, with its many tech-based
and women-led companies, Michelle says inspiration was all around:
"Lots of moms and dads around me had interesting scientific and technical careers. Even if some of the women had opted to go home to care
for kids, many of them had already had pretty high-powered careers,
so that for me was the norm. I didn't really see anything different until
I broke out of my bubble and saw this wasn't every woman's experience, to walk into a room feeling like an equal."

tradeoff between planetary boundaries and raising the economic bottom rung of humanity," says Michelle. "It's immoral to prioritize one over the other. It's incredibly nuanced and really challenging, but it's one of the reasons that working on Biomilq, specifically, is the dream. We get to thread the needle in thinking about how we solve these issues without having to choose between people and planet."

"Biomilq is kind of the perfect opportunity to talk about sustainability without having to sell someone on sustainability," Michelle says. As she describes it, it's a product that's healthier for babies and will empower women's lives, "and also happens to be better for the planet." Rather than just putting products out there because they're more eco-friendly or sustainable, we need to be developing products that actually meet a need and solve a problem, Michelle explains. That's true sustainability. "We're solving a fundamentally challenging issue for basically every woman on this planet," she says, "and it happens to protect the planet for future generations to inherit."

When the time came to start raising money, like most first-time founders, Michelle and Leila did a lot of learning on the fly, not only figuring out how to formulate an effective pitch but also strategizing who would make for the best partners in the long term. Early on, the founders realized they didn't need to feel tied to approaching only the typical players in food investment. "We really wanted to find investors that we felt would support us, and understood the kinds of issues we were trying to solve, and believed in the technology and the potential for it," says Michelle. "We started to see which funders would be good allies and were interested in the possibilities." Michelle says it was evident pretty early on who got it and was aligned with their mission, and who wasn't a good fit because they were looking for a shorter-term investment. In the end, the pair ended up concentrating on investors who were focused on sustainability, women's health, and health technology, rather than food agriculture and technology.

As Michelle sees it, evolving our biotech and food-tech systems to offer greater access, equity, and sustainability will shift the hierarchical structure and power structure currently present within the systems. Michelle emphasizes that it's important to bring this kind of engagement to every sector, even and perhaps especially those that are less flashy or not visible: "If we don't have institutional change in all of these sectors, it doesn't matter how cool the tech or science is—it won't create the change that we as founders want to see."

When it comes to being a woman founder, Michelle says that entrepreneurship provides tremendous opportunities for women who've faced gender-based roadblocks in their careers: "I think female scientists, especially female technical leaders, have always been looking for more meaning in what the science field can do, because they don't always fit in with the grind of being in a male-dominated field." Michelle says it can get tiring being the only woman in every space where you work, whether it's the research bench or advocating for new lines of scientific investigation: "There is some level where you start to say, 'Okay, well, if I'm always going to be on the outside anyway, what do I care about? What does science mean to me?' and you focus on the other applicability of your technology." She underscores that many women are interested in their work having a broad social impact and elevating the quality of life of people around them.

This echoes my own experience talking with a variety of women in tech and other STEM careers, many of whom have faced gender-related barriers. It's incredible how many women press forward despite such resistance, but, as Michelle notes, that could be in large part because so many women are especially motivated by more than just making money—they want to make a real, lasting, positive change in the world, and that keeps them plowing ahead. Even among those in more traditional roles in Fortune 500 companies, women are frequently part of affinity and social-action groups, and in other initiatives aimed at changing workplaces and work products for the benefit of all. No matter where or how you work, there are a million ways to make a difference. "It's becoming more and more normal to bring your passion and beliefs and cares to the work you do and to try to find impact in everything you work on," she says.

When I ask Michelle the biggest mistake she's made so far in her career, she says it's been taking things too personally, like when a pitch didn't go particularly well: "I am my harshest critic. I knew that going into this, and one of the skills I'd been cultivating was learning to let go when I make mistakes or when I perceive that I've made mistakes, but I'm still not great at it." The reality, she says, is that it's a team effort, and realistically you can't solve everything by yourself. It's more effective and healthier to look at those incidents where things didn't go well and see where you can improve, then move on. Michelle says that as a younger founder, she's been prone to doubt her capabilities, yet, as she notes, "The truth is, nobody really knows how to do everything."

As a first-time founder, Michelle has a remarkable amount of self-awareness. While she says she has lots of passion and compassion, being a warm and fuzzy leader isn't her strong suit—she's more of a high-level thinker and strategist. Michelle says that being gentler with herself and others is something she's working on and is an important skill set for her to develop, but she's also conscious of the importance of playing to her strengths and not trying to be someone she isn't. It's the type of balancing act lots of women in the business world have to manage: "I think, naturally, especially as women who want to see this dramatic change, who want to create this new world, we're always hustling. We're always trying to get everyone to be the best they can be to create this utopia we're dreaming of. Seeing some of those areas that I need to tilt toward and develop is important, but we're also successful founders in part because what we strive for is creating that world we want to see."

Chapter 5

Lisa Feria

CEO and Managing Partner of Stray Dog Capital

Building a Refugee's Legacy, Disrupting Venture Capital, and Paving the Way for Latina Leadership

"When we close our eyes for the last time, what are the things we want to be proud to have done?"

When it comes to the world of entrepreneurship, it's easy to focus on founders—yet without funding, no matter how world-changing their ideas are, many will never get their companies off the ground and scale them to success. That's where venture capitalists like Lisa Feria come in. Lisa is the managing partner and CEO of Stray Dog Capital, one of the world's first venture-capital firms focused on the plant-based and animal-alternatives sector with early investments in some of the biggest brands in the industry, including Beyond Meat, Miyoko's Creamery, and Good Catch. For Lisa, her work stems from her passion to make the world a better place for generations to come, and that passion is fueled by the generations of family members who came before her.

Children of immigrants are often raised with a deep sense of appreciation for opportunity and frequently adopt their parents' drive to succeed. Lisa Feria knows this firsthand. While Lisa spent her formative years in Puerto Rico, her family's story starts somewhere a little bit farther north: Cuba. As refugees of the Castro regime, Lisa recalls how her family made the brave decision to flee for a better life. "If you left Cuba during that time, you only could take what was on your body," Lisa says. "No jewelry, no money. As soon as you walked out of the house, anything you had was taken by the government, so you left with nothing. My great-grandmother wanted to give her family a shot, so she very courageously sewed small valuables into her belt even though she could have been killed if she had been caught."

From Cuba, the family went to Spain, where they didn't know anyone, and Lisa says that while there they relied largely on the kindness of strangers to get by. Eventually they made their way to Puerto Rico, where Lisa's father met her mother. About her family's early days, she says, "There was a lot of restarting and re-rooting and reseeding and reinventing themselves in many different ways and shapes, and I think part of that spirit has really carried on to me. In my father's family, there is no such thing as being jobless. There was no excuse of 'Oh, I didn't get trained in this' or 'I didn't get educated in that.' It was such an entrepreneurial culture." Lisa says that in her experience, the Cuban culture in particular is very entrepreneurial: "My grandmother started out selling whatever she could and eventually ended up having her own store."

I can personally attest that for folks coming from an immigrant background, the "immigrant hustle" is real. Hailing from a family of immigrants and being an immigrant myself, I know there's a certain sense that is bestowed in you to make the most of your adopted nation. Immigrants to the US, for example, account for a vast number of the nation's entrepreneurs. In particular, during my career in the tech industry in Silicon Valley, I found myself constantly surrounded by newcomers from all around the world. While the media's typical depiction of tech founders is White male founders, indeed, more than half (52 percent) of tech founders in the Bay Area are immigrants.

Using the money Lisa's grandmother made from her store, she was able to send Lisa's father to school, making him the first person in the family to graduate from college. That was no small task, given not only her father's humble beginnings but also the fact that while attending college, at age eighteen, he became a father. Lisa's mother became pregnant when she was just seventeen: "My mom has definitely been one of my sources of inspiration for just being incredibly scrappy. Not being afraid of facing really big, audacious, hairy challenges, and just leaning forward and trying to do your best and sort of jumping." Lisa's mother worked several jobs, including spraying perfume in a department store and running a makeup counter, patching things together until eventually she started her own business. "The women in my family are badasses," Lisa says. "My family has very strong women—strong men, as well, but it wasn't one at the expense of the other."

For Lisa's part, her path to leading Stray Dog Capital started with a childhood ambition of becoming a vet: "I love animals. I'm a vegan. My family's vegan. We have such a big heart for animals, and ever since I was a little girl, that's what I wanted to do—to save animals and make them not hurt." Eventually, however, the idea of doing something that felt safer and more mainstream won out against the risks of entrepreneurship and starting her own company, so at school she studied chemical engineering. "I wanted to have a job where you checked in and checked out and that gave me a great paycheck," she explains. "Chemical engineering was very mentally and academically difficult, but also stable."

Out of school, Lisa landed at General Mills and immediately found herself supervising mostly men, pretty much all of whom were considerably older than she. "I had a fast, painful course on conflict management, relationships, and being very humble," she says, "but I also quickly realized there that I wanted to do more." She went back to school and earned her MBA, then took a position at Procter & Gamble: "Both of those companies were the definition of mainstream—check in and check out, safe, great benefits."

Lisa loved her work, but then, as happens for so many of us, she had a life-changing interaction with animal agriculture: "PETA crashed into my life when I saw the video *Meet Your Meat*. I had a major moment of 'this can't be right.'" When she did some research, Lisa discovered that confinement-based agriculture had become the norm, far from the common conception consumers are given that the animals responsible for their food are whiling away their days in beautiful pastures. "When you start treating animals like widgets or tools, then you have all this horrendous treatment," she says. "I had a big come-to-Jesus moment. I looked at everything else in my life.

I became vegetarian overnight." After seeing *Forks Over Knives*, she went vegan, but for Lisa, that still wasn't enough: "I decided I needed to make a jump here. I needed to do something that matters, because I don't want to hand this world on fire to my children. So, at that point I said, 'Okay, I need to go to the private market and do something entrepreneurial. Because I believe that if you have products that people love, they will make the easy choice and pick them.'"

Lisa was invited to interview with Stray Dog Capital, which was based in Kansas. Though the mission lined up with what she wanted to do, Lisa was convinced there was no way she was moving from Cincinnati, where Procter & Gamble was based, to Kansas. "I'm from Puerto Rico," she reiterates. "In Kansas, you can't get any farther from the ocean. But then I met the founders of Stray Dog Capital, and I said, 'Yeah, we're moving to Kansas.'" What she saw was the potential to rebuild the entire food system, so in 2015, she joined Stray Dog Capital as CEO.

I ask Lisa if she believes that being a mother has influenced her career trajectory. "Absolutely," she says. "Up until I was a mother, I was thinking about ways to maximize my career and support our family. That was really all the math I was doing. When I became a mother, all of a sudden I started thinking about a longer timeline. It was no longer about what was good for me in my lifetime, but I cared about multiple lifetimes and multiple generations after mine. It didn't matter how many paper towels I sold or how much money I made, because we are in such a crisis from a climate-change standpoint. We have driven ourselves into such a terrible situation with animal agriculture; human health is in a really bad place because of the ways animal products affect our bodies. I saw all of these things and I saw that they all added up to a situation that would put the world in peril for generations."

Lisa says this is her life's work: "The more people that I can support and inspire and encourage to come into this category the better. It's a many-hands, many-legs, many-brains situation. We need as many people as possible to be working to improve the food system. We need all the money and all the talent to be working toward that problem."

The role of venture capital is often misunderstood or stigmatized as a ruthless business aimed at siphoning value from hardworking founders and maximizing profit at all costs. Indeed, there are countless horror stories and many bad investors out there, but Lisa is a leader cut from a different cloth entirely—one focused on mission, impact, and changing the world. Recalling the first pitches she heard, Lisa shares, "I thought to myself, 'We are empowering the angels. We are empowering the people who are going to take these great ideas and put them into the market and save lives and make people healthier. We want to be the group that continues to empower those amazing, talented people who are doing incredible, inspiring work.'" But while investment is critical, Lisa insists that Stray Dog Capital and other investors play a relatively small part: "At the end of the day, the entrepreneurs are the stars of the show, and we are very much the supporting cast."

Lisa and Stray Dog Capital were in the game at a time when few were investing in plant- and cell-based companies. At the time, much of the investment focus was on mobile apps and similar digital technology: "There weren't a lot of investors. Fast-forward to Beyond Meat's IPO, when all of these people realized that this area is super exciting. We at Stray Dog Capital were like, 'Yeah, we know!'"

These days, funding for such companies abounds, and Lisa advises entrepreneurs to proceed with caution and consideration: "If you're a smart, strategic entrepreneur, you can find money. What's more important is finding the right strategic partner who can help you get to the next stage, who has the right contacts, and who can support you in follow-on stages. It's like finding a marriage partner. You have to find someone with the right qualifications." Instead, Lisa says, when there's loads of funding available but a dearth of entrepreneurs with great ideas, companies can accept "dumb money"—money that comes without that critical support from investors who just want to turn the fastest profit possible. They can pour money into a company too fast, too soon. Especially for food companies, which take some time to get a product to market and therefore generate revenue, it can be harder to generate capital later when they really need it to scale up and build out a distribution system. At the end of the day, the company could collapse. Instead, Lisa says, "Make sure that the money you're taking is valuable and your valuation makes sense," otherwise everything you've worked to build could burst like a bubble.

In the past decade, nearly $18 billion has been invested into animal alternatives, such as plant-based and cell-based technology—with a staggering $6 billion in the year 2020 alone. Lisa says that even though there was a balloon of investing during the past few years and some companies that she believes shouldn't have been funded received investment, plant-based food products "are still a sustainable place to invest and get great results." For Stray Dog Capital's part, Lisa says the firm is "laser focused" on funding products in areas where current offerings are most environmentally harmful, and that have the broadest market. These include categories such as meat, seafood, and cheese, as compared with products such as nutrition bars or soups, which no longer garner interest from the fund.

Stray Dog Capital also is working to create lower-cost plant-based alternatives: "We can't continue to have products that are one and a half to two times as expensive as the meat alternatives. We have to find cheaper ways to make these products." Looking to alternative protein sources beyond soy or pea protein to a bigger variety of possibilities is part of this: "It's hard to find companies that are using other proteins, so that's a focus for us." They are also looking at mycelium-based (fungi) products, which could represent the next generation of plant-based meat products that taste as good to omnivores, if not better, than the animal meat they're used to.

Another of Lisa's powerful personal aims is to elevate women and other underrepresented groups, especially Latinas, in the world of entrepreneurship. She says she feels a tremendous sense of obligation not only to her family but to other Latinas to be successful: "You really do feel like you have a responsibility to prove that we deserve to be here, that there should be more Latina venture capitalists, and to open that door wide so that all the people behind me can come through. I absolutely abhor hearing that there are only a few of us."

Lisa recalls attending a conference and learning that, at the time, the number of Latina venture capitalists who lead their funds and are managing funds of $50 million or more was only five. Lisa says, "I turned to the person next to me and said, 'Did she say five percent?' and she said, 'No, five.' I was floored. I knew it was very low, but I never knew you could count it on one hand." In 2021, Lisa was on a panel that aimed to inspire future generations of Latinas into venture capital, and she learned the number had increased . . . to eight.

Turning to founders, the numbers remain bleak, as well. In 2020, women founders received less than 3 percent of all investment world-wide, with women-of-color founders raising less than a single percentage. With record investment in the animal-alternatives space, unfortunately, this funding trend is also playing out in much the same way. Through my organization's annual founder survey, we spoke to 160 women founders and CEOs and learned that not only did nearly half (48 percent) of all founders report experiencing bias in fundraising, but nearly half (47 percent) of all founders of color experienced racial bias, specifically from investors. There is still much work to be done.

The work of diversity and inclusivity not only among entrepreneurs but also among venture capitalists is so important, Lisa says, "because if you don't see anybody that looks like you on the other side, this is where imposter syndrome becomes a thing. When you're working so hard to make it and look around and see so few people like you, you begin to wonder if you can make it. You really have to keep your eye on the ball in terms of not letting that mess with you. In so many of my fundraising presentations, I will be the first Latina that they have ever talked to for a fund of our size. I don't want to be the first one. I'm grateful to be here, but I wish that we were at a point that I'm not the first in anything. But here we are. We need to fight to make a difference and make it better. You may not be able to look around and see a lot of people like you in these positions, but that doesn't mean you can't make it, and that doesn't mean you can't make it better for others coming behind you. The more people I can inspire to come into this category, the better."

To help make it better, one of Stray Dog Capital's primary objectives is to support people of color, especially Latinx people (people of Latin American heritage), in venture capital. They prioritize hiring Latinx interns and referring potential employees to other funds, and they've started a Latinx venture-capital program at a local university. Personally, Lisa takes every opportunity she can to amplify the achievements of other Latinas, including shout-outs on her social media platform. "I think women are not great at amplifying each other," she says. Lisa says that over and over again in boardrooms, she sees others freely amplifying one another's ideas and insights. When you get two or three people saying, "Mike made a great point there, and this is why," suddenly Mike's idea carries three times the weight and momentum: "Amplifying other women and people of color is so critical, first because we have to continue to inspire others to enter these fields so we don't have any more firsts. There shouldn't be eight anything. We should have hundreds of Latinx women and men in these positions. Second, because it's up to us to impact our own circles as much as we can. Each one of us does influence and control our own circle. If each of us can make a ripple in our own circle, together we can make a wave."

When it comes to overcoming imposter syndrome, Lisa's view is that you'll always have that critical voice inside you trying to hold you back—you can't get rid of it entirely. "But what you can do is make that other voice stronger," she says. It's the voice that encourages you and cheers you on, and if you learn to listen to that one more, to amplify that voice at your own internal boardroom, you'll make it.

For her part, Lisa is using her own powerful voice to help and encourage all of us to eat differently. She says, "We have so much individual power in the choices that we make, from the food that we eat to who we follow and amplify, from calling representatives to using our power to help others who haven't had the same opportunities to get ahead. At Stray Dog Capital we are a small organization, but we still have the ability to mentor people, to hire people at all different levels." Reflecting on her role, Lisa holds a sense of obligation much deeper than a simple professional one: "We all have some measure of power and strength, and it's so important to use it. We have only one life. When we close our eyes for the last time, what are the things we want to be proud to have done?"

Chapter 6

Shama Sukul Lee

CEO of Sunfed

Facing Existential Crisis, Engineering a New Way to Nourish, and Rebalancing Capitalism with Feminine Power

"It's not time for change; it's time for transformation."

Shama Sukul Lee started as a software engineer who, after a life crisis, decided to turn from digital technology to food technology—despite having no background or skills in the food industry. Her company, Sunfed, uses proprietary processes and custom hardware to make lean, clean proteins to replace animal-based proteins in the global food market. Based in New Zealand, Shama and her company have emerged as an early leader in the Oceanian (Australia and New Zealand) plant-based industry. Shama's view of business is a unique one, focused on creating conscious consumerism to replenish, nourish, and restore feminine energy in our modern industry ecosystem.

"Sunfed was born out of an existential crisis," says Shama Sukul Lee. Shama's story is that of an unlikely CEO. As a result of excelling in academics, Shama, born in Fiji, received a scholarship to study in New Zealand. After she graduated with an MBA, she became a software engineer, found her partner, and settled down, or, as she describes it, "checked all the boxes that are supposed to make you happy. But I was beginning to feel hollow and unfulfilled and disillusioned with life," she says. It was a turning point for Shama, who quit her job and put herself in what she describes as a kind of self-imposed exile for a year, where she separated from social media so she could get a sense of why she was so unhappy.

The result, Shama says, was the realization that she'd been "living on autopilot," wearing the various identities she perceived society wanted from her without really considering what it was she wanted to do and who she wanted to be. "I shed all of that," she says, "and it was a very uncomfortable experience because you lose yourself. But I'm very grateful it happened, because only then did I become really conscious of my life and the fact that it's finite. We're here on this planet with a limited amount of time and energy, and only we can decide where we want to put that time and energy," she says. And it was out of that experience that Sunfed was born.

At the time that Shama stepped back from everyday life, she was also on a journey to eat less meat, and eventually it dawned on her that the food-sustainability issue was one that she could take on. "Food has become many things," she said during a keynote address for the City of Sydney Visiting Entrepreneurs Program. "It's a source of pleasure, a source of comfort, a skill, an art, a social activity, an addiction, or just something to do when there's nothing to do. Fundamentally, though, food is fuel. It's the largest energy we consume on this planet, and the largest market with the biggest footprint on this planet. And the largest part of that footprint is made up of animal protein." Shama says that the inefficiencies and outright deleterious effects of relying on animal-based sources of protein—from the feed and water consumed by animals to the need to dispose of their waste to the massive amounts of pharmaceutical products required to keep animals in high-intensity operations relatively healthy—all add up to a strain on resources, along with increased pollution of soil, water, and air, deforestation, and habitat destruction. "Meat and dairy," she says, "are the fossil fuels of the food world."

Conversely, as Shama explains it, the closer we get to the source of our energy—the sun—the cleaner and more efficient our fuel becomes. Since humans don't yet have the ability to photosynthesize, the closest we can get is to skip animals and instead eat plants. Shama draws inspiration from the sun, the namesake of her company, and its natural ecosystem. "Plants are the original solar cells and the original battery packs," says Shama. "They capture and store energy into a form that we can then use."

Long before COVID-19, Shama was also focused on the issue of pandemics. She underscores the additional threat to humans posed by viruses that can jump from animals into the human population, something that is far more likely to happen when animals are raised in massive quantities in close quarters, which is the situation on factory farms. Today, an estimated 99 percent of farmed animals (excluding fish) in the US live on factory farms, with roughly 90 percent of farmed animals worldwide raised in these conditions.

However, in spite of these clear drawbacks to animal consumption, consumers continue to clamor for animal products. One of the main challenges, says Shama, is that they're hooked on the taste. Sunfed has set out to create products that are not only delicious but also better for the planet in every way, and produced at economies of scale, so they're accessible and affordable for everyone. It's the last part, Shama says, where we seem to struggle, because "achieving those economies of scale is its own beast." The solution, as she sees it, is to be "super conscious about how we scale." Classically, when companies scale, one of the things they magnify is their deleterious effects, whatever those may be. Conversely, Shama's vision is to be able to scale in a way that has the opposite effect: restoring and rejuvenating resources instead of exploiting them. That's why Sunfed has elected to focus on pulses as sources of protein—dried seeds of legumes, which include lentils, peas, beans, and chickpeas. Pulses are considered one of the most environmentally sustainable crops because they fix nitrogen in the soil, don't require fertilizer, minimize pesticide use, have low water consumption, and tend to be drought resistant.

The way Shama sees it, companies such as Sunfed, which strive to leave the world a better place, will help to bring a balance of feminine energy to a business world that's been almost entirely dominated by masculine energy. Shama's quick to clarify that she's not speaking of women or men per se, but rather of how we view resources and the approaches we take toward business. Rather than overfocusing on industrialization, cold logic, conformity, and mechanization—a system where resources are exploited for profit—this shift will focus on more-sustainable methods that work to not only maximize but actually renew resources. "We have to stop redesigning the systems we've inherited and instead design ones that will actually rejuvenate us," she says. "It's not time for change; it's time for transformation. Sunfed isn't just in the business for profit; it's in the business of making all life better."

For Shama, it's not about a complete shift away from what we've done before, as we still need many of our masculine qualities to dig in and build businesses, then scale them. Instead, it's about taking a more balanced approach in which we consider impact over profit: "For Sunfed, we're very conscious that we don't want to become another problem on the planet. We've seen companies start with good intentions, and as they grow, then they start exploiting, using too many resources, and so on." And avoiding that fate, says Shama, requires constant tending to every aspect of your business.

"Your company is a living, breathing thing. It evolves. And some people who were well suited for one stage are not well suited for another stage. With Sunfed," Shama says, "we find that it outgrows us really quickly. We have to keep up, and we have people who enjoy that because it keeps life interesting." Shama was used to this from the start, having transitioned to creating the company directly from working as a software architect with no experience in the food industry. Yet she says her and her team's inexperience turned out to be tremendously beneficial, because "we weren't limited by the existing paradigms in food." Yet, she adds, the transition of going from a very cerebral space to one that deals in the actual manufacturing and physical consumption of products was very humbling. As she describes it, it was a shift from dealing with bytes to dealing with bites.

Creating new products meant literally creating new technology, as Shama and her team found that to accomplish their goal, they really needed to create their manufacturing systems from scratch. Shama says it was a challenge, but, "I tell my team that we're on the bleeding edge, and when you're on the bleeding edge, you're going to bleed a little bit." As she told *Vogue Australia*, "To create software, all you need is coffee and a laptop. With hardware, each iteration or prototype requires significant capital. And there is no plug-and-play infrastructure;

we've had to build all that up from scratch. Starting a company means you're constantly being stress-tested," she adds. "There is one challenge after another, and you have to learn to enjoy the process, or you'll crumble with the pressure and fast pace."

Part of that process is scaling. In 2018, New Zealand–based Sunfed raised NZ$10 million in a Series A fundraising round. When it comes to how to choose the best investors for your company and make sure you don't get into a position of losing control, Shama says it's important to remember that only we can give up our own power. "To the entrepreneurs out there, always remember that you, the founder, are the creation engine. You are upstream—everything else is downstream of that. Without you," she says, "this thing doesn't exist, so there's your power. Investors need you more than you need them, especially now in a capital-fueled world."

Shama notes that while it's important to choose investors who are a good fit and to nurture those relationships, "you need to keep control of your company for as long as you can. For me," says Shama, "I have a really big vision that I want to see become reality, and I don't want to compromise. At the end of the day, no one will care about the company and the vision as much as you do."

That said, Shama explains, your vision will grow and shift as you do, and you also want to be involved with investors who allow those shifts to happen rather than have to have every detail mapped out in advance. "When you're looking for investors, you need to understand your term sheets right from the beginning," she says. Shama also underscores something that Slutty Vegan CEO Pinky Cole (chapter 2) preaches: be on top of your financials from day one and have excellent legal representation.

When it comes down to it, though, Shama says that if you stay focused on your mission, in time you will achieve your goals. It's important to get the details right to make the business work, but it's your big picture that will make your company's success the world's success. "Be in the business of making life better," she advises. "If you're in the business of well-being, then everything will align and the problems we're currently having will be solved."

Chapter 7

Fengru Lin

CEO of TurtleTree

Leaving a Dream Job at Google, Taking a Systems Approach to Food, and Shaping the Future of Human Milk Products

"... in many ways, TurtleTree isn't offering
an alternative protein but 'the original
protein we were meant to consume.'"

What would be compelling enough to make a successful young tech professional with a dream job at Google switch direction into totally uncharted territory? For Fengru Lin, it was a call to make a more immediate, lasting, positive impact on the world around her—to apply tech to the future of food. As the CEO of TurtleTree, Fengru and her partners are poised to disrupt the dairy industry, creating safer, sustainable, and delicious options using cell-based technologies. But that's not where Fengru's vision ends—looking to create an entirely new industry altogether, she plans to also license TurtleTree's technologies to produce sustainable and humane nutritional products to companies around the world.

When Fengru Lin grew up in Singapore as a child, the small island was far from the emerging tech powerhouse it is today. With less than three hundred square miles to produce food for five million people,

Singapore is a country that relies upon calculated, precise action and ingenuity to survive—and thrive—as a nation. Indeed, Fengru believes that the exceptionally structured, solution-oriented environment not only felt good to her as a child but also inspired in her the mindset that with careful thought and directed effort, even the most complex problems can be solved. "For me, I always had the feeling that everything was going to turn out fine," she says, pointing to the example of Singapore's organized effort several decades ago to become more self-sufficient with its water supply. As a small island nation, Singapore has no natural access to fresh water. In 1965, the young country launched a massive, multidecade organized effort dubbed the "Singapore Water Story" and created two desalination plants and seventeen reservoirs, installed five thousand miles of waterways, including drains and canals to funnel water to the reservoirs, and developed water cleaning and recycling systems. To this day, Singapore remains one of the only countries to have such cutting-edge water technology.

"It was a major success, and today we even export some of the water we process," Fengru tells me. "Seeing this kind of thing made me appreciate that when we set a goal for ourselves, we can do it." As we chatted together during the peak of wildfire season amidst miles of endless dry rolling hills in California, it seemed downright shocking to me that a country could export water. It was also apparent in the way Fengru spoke of her home country that she held both a reverence for its power and a certainty in its future—a special confidence that would go on to shape her perspective and innovation as a founder.

At a time before climate consciousness became the norm, Fengru grew up with an awareness of the need to consider the environmental impacts of her actions from her earliest years. When Fengru was younger, beyond its water system, Singapore didn't do much in the way of recycling. Fengru's mother is Taiwanese, however, and when they would go to visit her mother's family in Taiwan, recycling was the norm there. Fengru remembers receiving strict instructions about how to separate out the trash so it could be recycled. These seemingly small actions would go on to shape her worldview.

Fengru further credits her love of the outdoors, including hiking, cycling, and snowboarding, for her awareness of environmental concerns, noting that the snow-sports season in some areas around the country is shorter by a matter of months due to climate change. It was actually this increased impact—and subsequent awareness—of climate change that would make Fengru shift the trajectory of her work from a traditional tech career to focusing on technologies that could improve the quality and sustainability within our food systems: "The tech space is still interesting, but for me, the call of food and sustenance is what's important, and the tech can support that."

At school, Fengru studied information-systems management, primarily centered on business. Early on, she worked for a two-hundred-person Dutch startup with a small family-like atmosphere, which was in the process of making the leap to becoming a major corporate player. From there, Fengru launched into a career with massive, industry-leading companies, first with Salesforce, then with Google. From these diverse experiences, Fengru says she got a good sense of the structures that make companies work well at scale, as well as how to balance that structure with the flexibility necessary to pivot—experience that serves her well now that she's running a small multinational company of her own. Pairing her can-do mindset with her tech background and practical business skills, Fengru began to develop a vision for how she might shift her career to be more impactful.

The idea for TurtleTree came about in a rather likely place: Google. During a presentation on transformative technologies, Fengru learned about food technology, specifically cultivated cell-based meats and Uma Valeti's Memphis Meats (now Upside Foods), for the first time from her now cofounder, Max Rye, who was then the CEO of another tech company. "After the talk we started chatting," Fengru says. "This was a time when I was really a fanatic about cheese-making. It was so crazy I even went to Vermont and upstate New York for a few months to learn about how to make cheese." Fengru recalled how the cows there looked so happy and how they grazed freely in the fresh air. "At the time," she recalls, "there was no good source of raw milk in Singapore."

Upon digging further, Fengru quickly learned that the idyllic concept of milk production she had seen in Vermont was far from the reality of how nearly all dairy in the world is produced, and how nearly all cows in the world are treated. In addition to learning about the inhumane methods of animal confinement, she found that the dairy industry faces challenges around contamination with antibiotics, hormones, and bacteria, an issue seen across the animal-agriculture industry internationally. To give some perspective on how we have pushed biological limitations when it comes to dairy-milk production, in 1934 a typical cow produced 1.2 gallons of milk daily. In 2017, that number was up to 7.2 gallons. It's similar to how technology and breeding practices have been used to produce chickens and other birds for consumption whose bodies are so unnaturally large, their legs can't support their weight.

Fengru says that current industrial farming methods are not only bad for cows but bad for the milk, which doesn't have the balance of ingredients needed to make the high-quality dairy products she was aspiring toward as a hobbyist cheese maker. Talking with Max, she saw an opportunity to make a product that was safer and more humane, sustainable, and nutritionally beneficial: "So, we started chatting about how to use these kinds of cell-based methods to make milk. Back then we were making some of the proteins in milk, but no one was making actual milk."

As Fengru explains, and as echoed by Biomilq founder Michelle Egger in chapter 4, milk is an exceptionally complex product. For one, there are two thousand different ingredients. You read that right—two thousand: "Being able to make butter, cream, different kinds of cheeses—it all starts from milk. It was fascinating to me that this one starting point could result in so many different food products. The whole process got me to appreciate how complex milk is, and how we really needed to access not just the proteins but the whole variety of ingredients in milk."

To Fengru, the possibilities were endless. Her parents, however, didn't see things quite so clearly. "When I told my parents what I wanted to do, my mom was freaking out. She was crying and asking what was wrong with my job I already had, how was I going to have enough money . . . I had to assure her that things were fine, I had money in the bank in case things didn't work out, and, hey, if things didn't work out, they had extra bedrooms and I could stay with them," she jokes. When it came down to it, though, they supported her decision to start TurtleTree. Fengru says that in many ways, her upbringing was traditional in that her mother and most of her friends' mothers stayed at home to raise them, and neither of her parents was an entrepreneur. Yet, she adds, "Singapore is a very equal-opportunity space. I used to study at a girls' school, too, so I had a pretty good experience coming into the entrepreneurial space."

As many entrepreneurs do, Fengru decided not to leave her regular job right away. But after some exploratory science, in 2019 the scientific team Fengru and Max put together came up with several breakthroughs that prompted them to file patents. "That's when I knew it was time to leave Google and start TurtleTree," says Fengru. Most of TurtleTree's staff are based in Singapore, with a global research-and-development headquarters and a pilot facility in California, which they're staffing in conjunction with University of California, Davis.

Not only was the team propelled by its own early scientific success, but TurtleTree is also buoyed by a focus in Singapore on the next generation of food production. Similar to its Water Story, Singapore now has a Food Story, with a goal of producing 30 percent of their own nutritional foods by 2030. "The government is really all hands on deck to try to make it happen," Fengru says. "In addition to taking a visionary approach, we're also very pragmatic about it," she explains. "I was on a panel with one of the ministers, who said it's great that we can produce milk, meat, and bioreactors, but had we thought about all the materials that go into the bioreactors or the electricity that powers them? It's a fair question, because about 60 to 70 percent of the cost of producing foods in a bioreactor is related to the energy to power them. Where should that power come from? These are all things we need to think about when we consider what we need to do to become food secure."

I ask Fengru if, as a founder in her thirties, she's experienced any doubts or pushback from potential investors or partners because of her age. As she explains it, her advice to founders is to find a cofounder that has highly complementary qualities and skill sets, which, she shares, has been a huge advantage for her in dealing with high-level stakeholders. "My cofounder is American, and so it's like the world is his oyster," says Fengru. "He's also a big ideas man, and he can connect things really fast in his brain. For me, I'm from this structured background, and so it works really well. He comes up with the ideas, and I put them into boxes, charts, and lists, and we're then able to communicate it all to the rest of the team, investors, and so on." Also, before any big meetings, the two compare notes, which Fengru says helped her to feel more confident.

Further, Fengru explains that she had a fair bit of experience dealing with C-suite executives in her previous corporate roles, so working with heavy hitters wasn't new territory for her. "In this case, it was just selling an idea versus selling a tech product," she says. "Everyone's selling the future, and when it comes down to it, what matters is whether that path to the future can be clearly seen, and if you're meeting the milestones to get there."

While TurtleTree's first product aims relate to infant nutrition, it's not just babies who benefit from milk, says Fengru. When it comes to both dairy and human milk, there are some elements that provide essential nutrition for adults, as well, and may even have medicinal properties. For example, lactoferrin is a bioactive protein that is found in both human and cow milk, and Fengru explains that it helps to support gut-brain access and has anti-inflammatory and immune-boosting properties. Additionally, the complex sugars found in human milk are helpful for brain health, cognitive development, and gut issues such as irritable bowel syndrome. "At TurtleTree, we have a novel method to produce nineteen of these sugars using cell-based methods," says Fengru. "That's the deepest portfolio in the industry, so we can unlock a lot more functional benefits than elsewhere in the market today."

In the scientific community, there's speculation that if we consumed human milk (in place of cow milk) in some manner throughout our lifetimes, there could actually be sustained health benefits. Fengru says that for one, there are early reports that some of milk's complex sugars could help to prevent contraction of the COVID-19 virus because of their anti-inflammatory properties. There's also some investigation into whether components of human milk could prevent or help cure cancer and other diseases. "The proteins in human milk are a little bit different than those in cow milk, and the functional benefits we get from them are really like a lock and key. We can have trouble accessing similar benefits from cow's milk," she explains. As someone who hasn't consumed any dairy in nearly a decade for both health and ethical reasons, I felt that my mind was blown as Fengru shared her vision: What if we could move beyond creating plant-based dairy products for taste into creating human-milk-based products for health?

"The evidence is mostly anecdotal at the moment, but there's a lot of potential we can explore. But I think one of the reasons the research is nascent is the supply issue. Breast milk goes to babies," Fengru says. But if companies including TurtleTree can start making human milk, there will be plenty for scientific study. Right now, TurtleTree is exploring the use of cell-based human milk for athletic performance and as nutritional support for seniors. In fact, Fengru forecasts that their first product will be for performance nutrition.

"I think there will be a transition period, but as people start to recognize and understand and experience how some of these human-based ingredients and food products are better for them, they'll be willing to give them a shot," Fengru speculates. "People are starting to demand these alternative proteins, and there is definitely a shift in consumers' preferences. I see a lot of people around me, friends and family, seeking out better alternatives for the planet and for their health." As Fengru says, in many ways, TurtleTree isn't offering an alternative protein but "the original protein we were meant to consume."

Though TurtleTree has set milk as its first area of exploration, their vision extends much further. TurtleTree is also focused on producing the technologies—such as growth factors, scaffolding, and bioreactors—that will help companies innovate and scale throughout the cell-based space, resulting in more affordable and readily available products. "We've seen that a lot of these enabling technologies can benefit the whole industry—we don't need to keep them to ourselves," says Fengru. "I think that supporting the industry and also learning from them is something we want to do." As Fengru sees it, "We think TurtleTree will be everywhere. We think we'll have our technology in most food products, in some way, across the world in the next seven to ten years."

When I ask Fengru what advice she would offer new founders, she says that when she first started out, she would talk to "tons of people. I had calls lined up from a.m. to p.m., and I just wanted to talk to and listen to people, whether they were from the dairy industry or the food-tech space. Learn as much as you can, and speak to as many people as you can. And if you have a new idea, that's the best way to start executing one." Recalling how the support of others helped shape her vision for TurtleTree, Fengru firmly believes it is her responsibility to continue to help rising entrepreneurs, especially women like herself. When she can, she offers her time freely, remarking that founders are invited to contact her on LinkedIn. True to form, to this day, she makes time each Saturday to speak to would-be founders about their ideas.

Chapter 8

Riana Lynn

CEO of Journey Foods

A Journey to the White House, Serial Entrepreneurship, and Why Data Is the Secret to Healthy Plates

"Dare beyond your own doubts."

While Journey Foods focuses on a high-tech approach to food science, the beginning of Riana Lynn's personal journey with food was decidedly analog. Growing up, she began learning about food production alongside her grandmother as they worked the soil in her garden in Illinois. Since those days, Riana's career has taken her on a serendipitous and somewhat unbelievable path with a string of wildly successful cross-sector stints, including as an entrepreneur-in-residence at Google, a lecturer at Northwestern University, the founder of several successful startups, a White House team member during the Obama administration, and now her latest role as CEO of Journey Foods, a startup aimed at using data solutions to revolutionize the food system. Despite her high-powered tech focus, Riana shares why her passion still lies within the soil.

Riana Lynn was raised by a single mother, with help from her grandmother, in the Chicago suburb of Evanston, Illinois. Though she was raised in the Midwest, the South has been a heavy influence on her life and culture for as long as she can recall. Her grandparents migrated from Virginia, Alabama, and Georgia to the Chicagoland area between the 1930s and the 1960s, "when Black people were leaving farms to try and find jobs," she says. And their experiences helped to seed Riana's passion for food, health, and community.

One grandmother grew persimmons and watermelons on a farm in Alabama. Another grandmother, who was vegan long before the word and the practice were commonplace in the US, became one of the first Black yoga instructors in the country. "She exposed me very early to the ills of the food industry," Riana tells me. "She instilled in me this interesting sense of food and community, and food and wellness." Riana says she's grateful to have grown up with stories from generations before her of people becoming vegetarian or vegan and carrying their gardening tradition from Alabama all the way to Evanston: "All of my grandparents have a really big impact on how I look at food. Every day as I wake up, I think about how I want to leave my own mark."

Riana says that growing up in Evanston, she had the advantage of living in a progressive community, with a mother who made sure Riana had lots of early opportunities to be exposed to science, ecology, math, and nature. Also, thanks to her close proximity to Northwestern University, Riana was able to meet other students who were similarly interested in science. Though Riana was raised by her mother and grandmother, her father—an entrepreneur—was in her life, as well, and Riana says both he and her mother worked hard to put her in good schools so she'd have a solid foundation for her future. "I was just a nerdy little girl who loved basketball and science," she says. Because of her athletic prowess, Riana was recruited by Olympic track coaches at the University of North Carolina (UNC), where she competed in discus, javelin, and shotput, ranking as one of the top ten women discus throwers in UNC history.

At UNC, Riana went premed, with the idea that she could help combat the types of preventable chronic disease she'd seen so commonly growing up in Evanston, especially among African Americans. She explains, "Very early on I wanted to study how behavior and business and the food industry affected us." During her time in North

Carolina, she added African American studies (now Black studies) as a second major and took classes in computer science. Riana says that back before learning to code was common or cool, she'd be in her dorm room coding away, playing with graphic design and building websites. It was also in North Carolina that Riana was first exposed to a strong CPG and local-food scene. All around her were co-op grocery stores, small breweries, and locally made ice creams. It was then that Riana began to think in bigger and deeper ways about the power of local-food systems and their potential to influence health and wellness, and how technology could play a role in revolutionizing how we eat.

Yet in 2008, Riana also had another priority. "I just needed to work for the first Black president," she tells me. Barack Obama had just been elected president of the US, and Riana turned her sights to how she might land a role in the administration. She and a friend decided to create a media company, made their own press passes, then applied to "every event at the inauguration," she says. At one of these events, she met Dr. Rick Kittles, the cofounder and scientific director of African Ancestry, a company that traces genetic lineages of African descent. (You may have heard of them from their work with Oprah and other African American celebrities.) At the time, Dr. Kittles was conducting research at the University of Chicago, working in the same department as Michelle Obama. Not only was it a connection to the Obamas, but Riana was also fascinated by Dr. Kittles's research on genetics. She handed him a card and said, "I'd love to work for you." After learning more about Riana, he accepted.

While working for Dr. Kittles, Riana started attending grad school at Northwestern, where she studied genetics and biology, eventually earning a degree in public health with a focus on business and policy. During this time, Riana says she was also deepening her understanding of the challenges people in her Chicago community experienced trying to access fresh, nourishing food, along with how such lack of access contributed to chronic disease. At the time, she was living in a food desert—an area where fresh and healthy food options are hard to come by—despite being just blocks from one of the world's most endowed institutions, the University of Chicago. Riana wanted to shift her focus to influencing food systems directly, so in 2010, at age twenty-five, Riana created Peeled Juice Bar, a fresh-juice company based in Chicago. It was a crushing success, hitting $1 million in sales in its first eight months.

Though Peeled was rocking, Riana had not lost sight of her dream of working for the Obama administration. She asked Dr. Kittles if he'd be willing to write her a letter of recommendation to become a White House intern, which he did, and before she knew it, she was off to Washington, DC, where she worked on everything from Black issues to youth entrepreneurship. She even became a backup photographer for events. (You might spy her name in the photo credits among White House images published at the time.) "That was just an incredible experience," Riana says. "It was just thrilling."

Back in Chicago, Peeled was still growing. Riana knew the importance of scaling quickly and decided to create an e-commerce structure for the business. The tech behind online selling was no problem, but she had loads of challenges when it came to distribution. For example, how would she keep her juices cool for the two to three days it could take for them to reach buyers in Canada? Instead of starting from scratch to learn shipping and distribution, Riana turned to those who'd already solved those problems for their own companies—something she now advises other entrepreneurs to do. "Don't be scared to start a conversation and ask for the information you need," she says. "Turn to those you're striving to be like."

As Peeled's customer base increased, so, too, did questions about where the ingredients they used were sourced. As it turned out, these questions didn't have easy answers. But Riana agreed that everyone has a right to know where their food comes from, so she created FoodTrace to help track ingredients and answer these questions, marking her first deep foray into food tech.

Through FoodTrace, Riana discovered just how difficult it was to source organic ingredients at affordable prices. (Try getting local, organic kale during a Chicago winter.) This sparked a question that would define the next phase of Riana's career: Is there a way that technology can help make truly healthy foods more accessible to everyone? And by everyone, she meant everyone—not only people in Chicago's food deserts but consumers around the world.

It was around this time that Google came calling, offering Riana a role as an entrepreneur-in-residence. While she was there, she met Don Thompson, who had been the first Black CEO at McDonald's. Thompson had just left the fast-food chain and joined the board of Beyond Meat. He'd also cofounded Cleveland Avenue, a hybrid venture-capital and accelerator firm that supports innovative, technology-based food companies. Thompson asked Riana if she was interested in helping to direct strategy at Cleveland Avenue.

There, Riana got to consult and provide strategy to several multinational food companies and fast-growing startups before their IPOs. It was the missing insight she needed to come up with the idea for her next business venture—Journey Foods. In Riana's vision, the company would use technology and data to solve sourcing and food-quality challenges in the CPG sector.

Riana engaged several food and data scientists to enhance her initial software models that could improve development and save time on operations challenges for any CPG company, all using data models. With Journey Foods, she wanted to not only be able to help big multinational corporations (which combined are responsible for 90 percent of the packaged goods on our shelves) improve the quality of food distributed globally, but also to help entrepreneurs—especially underrepresented founders—scale their own food brands.

Journey Foods's JourneyAI technology analyzes companies' ingredients by nutrients and supply chain. Companies can upload data about their current products to get recommendations for how to enhance their products, such as using healthier and more-sustainable ingredients. The company then helps support the best ROI and manufacturing insights to answer the "then what" of their first set of insights.

As Riana said when I spoke to her, we often think about food as farm-to-table, considering only the farms on one end and only the restaurants or grocery stores where we buy it on the other. What we tend to overlook is what she calls the "complex middle"—the manufacturing side, which is where a lot of the technology and the cost comes in for companies. As she explains, "We want to make that part of the food value chain a little sexier and give more transparency to the complexities and the problems that exist, and also make it easier for brands to focus on what they do best, and that's market, and think, and ideate."

Riana explains, "Our grandparents and great-grandparents were not going to grocery stores in the same way that we are today. The earliest grocery stores were starting in America around 1947. We've had to accelerate the way that we develop food for millions of people in a short period of time. I think that process is messed up—food wasn't killing us like this sixty or seventy years ago. Our goal as a company is to unlearn the way that we've been developing food for the past few decades and not only reinfuse real consumer needs but also look for the gaps of what humans are doing when they're developing our snack foods and our sauces and our pastas and our candies and our sweets and our ice creams. Our team goes after the problems that exist in manufacturing those foods today; we're looking at what consumers' needs are, and we help companies make each product better."

As someone who's studied not only biology and chemistry but also genetics, Riana has unique insights into just how damaging today's standard diet is in terms of causing chronic health issues. She explains, "There are so many things we haven't considered in trying to create shelf-stable convenient foods over the past fifty years," including how people's genetic heritage impacts how they process foods and what nutrients they need. Riana says capitalism has taken us out of mindful manufacturing and research and development, with an overfocus on profit and speed and not enough attention paid to what's actually good for people. For example, it's common knowledge that eating a high intake of meat and dairy is bad for our health. "We weren't able to get where we are today and become thriving beings without the diets of our history, which were largely plant-based," Riana says. "Now we're on these new diets and we have chronic disease, we have heart disease, we have all types of cancers, all types of autoimmune diseases. We cannot thrive and build a country with three hundred million people *and* spend $300 trillion a year on healthcare or $300 billion a year on fighting diabetes." Yet, as Riana explains, roughly 70 percent of the calories that we consume each day are from these packaged goods.

By combining human thoughtfulness with machine learning, Journey Foods is looking to not only make foods healthier but make the planet healthier, as well. "What we've found as a company is that we really want to focus on the sustainability and the nutrition of these products," Riana says, adding that in many ways the promise of CPG has not worked out in either consumers' or the planet's favor. "Take the traditional methods for producing individual products such as fruit cups or fruit snacks, especially for the past sixty years, since grocery stores have launched. The cup's made in China, some of the fruit comes from Colombia and Thailand, the sugar is processed in another country, as are the graphics and packaging, and then they're brought together and shipped all over the US, and some of the products are even shipped back to where they came from."

A key part of the solution, as Riana sees it, is for more products to go plant-based. "I believe veganism is a huge part of the future in every community, but still, overwhelmingly, the food that we have created and manufactured today is not fully plant-based. And in order to meet the needs of billions of people, especially with ten main companies running most of the food brands and products around the world, we're going to need to accelerate and undo forty years of industrial change in our food systems, and so data is more important than ever so that we can reverse a lot of that."

Listening to Riana, it's hard not to be inspired about what's on the horizon when it comes to CPG. "What's most exciting about the products in the CPG categories is that everyone from every background and every socioeconomic sector is excited and learning and becoming more exposed to plant-based and better-for-you food. More people are excited about foods and ingredients that are less processed, that are gluten-free, and that create less water waste," Riana says. "For example, there are a wide range of African-native species that are naturally more nutrient-dense and organic, bypassing the GMO effects and poor soil quality in some areas of North America." She says the CPG industry is also looking toward scaling up the growth of Native American and North American indigenous foods that have been largely overlooked by industrial agriculture.

When I ask Riana what advice she shares with entrepreneurs, she doesn't hesitate. "Be daring," she says. Make the cold call. Publish the blog you're nervous to share. Make that press badge. "Also, find your tribe, because sometimes human capital is as important as that first $5k, $25k, or $100k. You'll get to those milestones, but if you find those people who will pick up the phone when you call and help you build a website, grow your social media platform, or answer legal questions, it will help you get there a lot faster. Resilience is built around the strength of your tribe, very early on, especially, and they'll help you dare beyond your own doubts."

Riana says that picking up a few extra practical skills can help during those times when you're waiting for checks to come in. In her early days, she was able to lean on her background in graphic and web design to help her span the gaps in cash flow.

Based on Riana's experiences, I'd add to that: be flexible. When Riana started out at UNC, she was premed. When she graduated Northwestern it was with a degree in public health. If she'd locked herself into one way of approaching her goal—of helping African Americans, in particular, get healthier—she might have become a physician or a public-health scientist. But, as echoed by Dr. Sandhya Sriram in chapter 15, Riana says, "I realized, you can be a scientific researcher or you can be a policy maker, but it's really the entrepreneurs that get things moving along the fastest."

Chapter 9

Heather Mills

CEO of VBites and Paralympian

Surviving Against All Odds, Choosing Conscious Capitalism, and Changing Diets Through Consumer Choice

"My whole ethos is: if there's a
problem, there's a solution."

Heather Mills has long been a trailblazer, but before she became a titan in the vegan food world, she was an inspiration to millions for her indomitable spirit. In 1993, while working in London as a model and advocating for refugees of the Balkan Crisis, Heather was involved in a collision with a motorcycle that resulted in her leg being amputated below her knee. As part of her healing from her accident, Heather switched to a whole-foods, raw vegan diet. Inspired by her plant-based recovery, Heather founded VBites, a vegan CPG company, which sells 104 products in more than twenty-four countries around the world. Incredibly, Heather also not only returned to modeling but became a champion Paralympic ski racer, setting the world record for fastest disabled woman in three sports.

Like many successful entrepreneurs, Heather Mills's solution-oriented attitude and strong work ethic were shaped in part by her early experiences in life. When she was just nine years old, her mother left the family. Four years later, her father was incarcerated. With no other choice, Heather started her very first business at age thirteen to care for herself, launching what would become a lifetime career in and passion for entrepreneurship.

"If I modeled my entrepreneurship after anyone, it was my dad, who thought he could achieve anything," Heather tells me, "even though he ended up getting himself in a lot of trouble. Hopefully I took the best bits of that and learned that everything is achievable." Heather says her father, who'd been in the military, had her on a tight schedule, getting up early, ironing shirts, and running the five miles to school and back because they didn't have access to transportation. "I suppose that gave me the discipline to do things and to be entrepreneurial," she says. When he was sent to prison, Heather left home and was living on the street. "It was a matter of treading on eggshells and learning what to do and creating different opportunities," she says. "Entrepreneurs can be born and entrepreneurs can be trained, and I think I was probably a little bit born, but mainly trained by the university of life."

Heather says that while she didn't have a stable family life, her friends helped to bridge that gap. From her childhood best friend, whose family fed Heather when she didn't have enough to eat, to a circle of about seven women she's been close with since she was young, Heather says her biggest supporters have almost always been women. "You can find inspiration in all types of women," Heather says, "not just those in business but those who are wonderful mothers, who are kind, who always have time for you and want to support you. I'd say I've gotten where I've gotten in the VBites business because I have four or five fantastic women around me who really listened to me. Now, in my business life, most of the strongest people around me in my business are women."

After several stints starting and selling small businesses, in 1990 Heather moved to the former Yugoslavia—now Slovenia—and became involved in the Balkan Crisis. As part of her work, she created a refugee crisis center in London, and it was there that Heather met with an experience that would alter the course of her life forever. A traffic collision with a police motorcycle left Heather with a fractured skull, punctured lung, broken ribs, crushed pelvis, and badly damaged leg. After the accident, Heather's leg was amputated below the knee. Incredibly, Heather managed to recover from her injuries, and was fitted for a prosthetic leg, though she was plagued by a persistent infection in her leg that would not fully heal. Desperate for resolution, Heather traveled to an alternative health clinic in Florida, where she was told that a raw vegan diet would help her. "For me, I was as ignorant as the next person about animal cruelty," says Heather. "It was purely for health reasons at first, but, like all vegans, you go for one or two of three reasons—either your health, or the animals, or the planet. But eventually you're vegan for all three, because you become educated."

Heather says it was easy for her when she first made the switch—but for her family, not so much. At the time, vegan meat replacements were essentially unheard of, so Heather set out to create her own to make the transition easier for others in her household. That was Heather's first foray into plant-based food production, and as a result, in 1993, VBites was born.

Heather started with a focus on soy, then became one of the early pioneers of pea protein. "We've been doing protein isolates from peas, lentils, chickpeas, and so on for a while now," says Heather, "but initially, they were expensive, and so the market wasn't quite ready for it." Conversely, subsidies allowed them to make soy-based products cheaper and therefore more accessible for consumers: "If you want to feed a family of eight for a certain price point and you want them to get off the meat, fish, and dairy, then soy is still the way to go because it's subsidized. You've got to make it cost-effective." Heather says one of the big challenges with soy, however, is some of the negative media around it, which has unfairly and inaccurately characterized it as problematic from a health standpoint. (Much of this is rooted in a misunderstanding of the difference between the human hormone estrogen and the phytoestrogens, or plant estrogens, found in soy.) "The problem is that the public doesn't understand the health benefits of soy," she explains, and that's part of the reason her company offers a range of options when it comes to protein. It's important to educate consumers, but also to recognize the reality that negative media, however inaccurate, "can bring a company down, and so it's important to have a backup solution," says Heather. "Nothing is black and white, so while we educate people, we also provide alternatives."

Today, VBites works with eight different proteins to create products for their various markets. Though soy is a fantastic protein, Heather says, the fact that it can't be grown easily in the UK means its usefulness in that market is limited. "Oats are the future here," says Heather, "as are algae, sunflowers, and things like that where we can create similar nutritious things we can grow here." For VBites, the UK is currently the fastest-growing market, though it's also the smallest: "The UK is absolutely booming. Before, you'd never see fast-food chains advertising vegan alternatives."

Decades ago, Heather also launched a number of vegan cafés, and she says the widespread adoption of vegan options among restaurants at every level in the industry is almost negating the need for them, except among consumers who don't want to eat at or patronize an establishment that also serves animal products.

"Everyone's got vegan options, and that's what we want, ultimately, so that the whole family can go together. When you give someone a bloody great vegan burger that tastes fantastic, eventually they change their habits," Heather says. "If you get them to reduce their animal consumption, they will change; people just need the time and tools."

Heather's work at VBites demonstrates the major role that CPG products are playing in the transition to greater consumption of plant-based food. As Heather explains and as echoed by Pinky Cole in chapter 2, getting people to try vegan alternatives to traditional fast food and similar products shows them that they can still have all the taste they want with plant-based versions. Eventually, though, as they're eating fewer animal products and their taste buds begin to change, they start to gravitate toward having more whole foods on their plate, as well.

Further, Heather shares, small changes in diet can have massive health benefits. In one program, VBites provided free plant-based meals to four thousand children in the Bronx, New York, for one year. "Just from giving them alternative meat, fish, and dairy products, we were able to reduce the need for insulin injections among those with type 2 diabetes," she says.

The view of changing the food system through conscious capitalism is an increasingly popular one. Conscious capitalism, a concept of business leadership where social and environmental impact is prioritized alongside profit, has gained momentum in recent years, particularly in conversations around the food system. Similarly, impact investing, the view of investing in businesses that prioritize this impact, has seen a rapid rise, with billionaire philanthropists, such as Bill Gates and Richard Branson, actively investing in companies building a better food system, such as Upside Foods (formerly Memphis Meats) and Beyond Meat. For many, impact investing, in which Heather is an active leader, poses a unique alternative or addition to traditional philanthropy. As we embrace a new generation of Gen Z change makers in the business world—the first generation to prioritize protection of the environment and sustainability as their first concern—it is likely we are seeing only the very beginning of the conscious-capitalist and impact-investment leaders to come.

Operating ahead of the curve, Heather's focus on the space may have come a bit early for some—including the market itself. Curious, I ask Heather why her early leadership in the space doesn't seem to have nearly the name recognition in some markets as newer brands like Beyond Meat or Impossible Foods. Though there are likely multiple factors behind this, Heather says that, as a woman founder, "I think you have to shout louder, though it may mean you're not going to be very liked. It's harder for women to be vocal about what they're doing and not come across as arrogant, and that's not just to men but to many women, as well." Having spent so much time in business and working as an activist and campaigner, Heather says she didn't think twice about advocating for herself. "I would just march into the boardroom and I would say, 'This is how it is, and this is what I'm going to do,' and I would just smash any walls down to make something happen." Yet, as she acknowledges, not everyone—women or men—is like that. "I grew up in my own sort of bubble," she says. "I was so heads-down that I didn't even notice there was prejudice toward women in business until about a year ago." Now, she says, she often travels to speak with different business groups and finds it's common for men in the room to not give women, including her, the time of day in discussions. "'We've run corporations,' they tell me, to which I answer, 'Yes, but you've never started a business and run it from the ground upward.'"

Heather says that because of such roadblocks, it's especially important for women to believe in themselves. "You've also got to get a thick skin," she says. Heather says that standing up for herself with media who were trying to slander her during her high-profile divorce (Heather was formerly married to Beatles member Paul McCartney) and taking up the cause of refugees helped to center her in herself. "You have to decide who you are and what you stand for in life, and just shout," she says, "but you're going to have to develop a hard shell, because lots of people are going to throw stones."

Additionally, Heather says, it helps for women in business spaces to not view one another as competition. "Women really need to support women more. And when you get your company sorted out, invest in other women's companies and in helping them scale. It's about helping each other." Heather's company VBites Ventures aims to create vegan companies that largely spotlight women, including a new cruelty-free cosmetics brand, Be at One, which Heather founded with a group of close friends.

A further piece of advice Heather offers is to make sure you actually want to be an entrepreneur. "Get to know yourself first," she says. "Entrepreneurs can't make a business without brilliant right-hand people. We wouldn't have gotten where we did without working with brilliant people. If you're not going to be the owner and the boss, you can still make a difference." A big benefit of working for someone else first, she says, is that you can start to learn the business without the risk and pressure. Then, if you're willing to take the risks involved, and your family's on board, go for it.

When I ask Heather what she sees as the area that has the biggest potential to expand the plant-based market—whether effective altruism, policy work, new products, or something else—she says, "I think it's all of it. Everyone has a role to play. Choose what area you want to be in, and if you have the tenacity and the time, choose more than one." Heather says that while she's always been an entrepreneur, she's also been an activist for a long time, and as part of her efforts to educate people, she's even ventured into documentary filmmaking.

Policy work, much like that of member of European Parliament Sylwia Spurek in chapter 13, is also key to Heather's business strategy. Heather says that as farm subsidies for animal products are being removed in some countries, VBites is working with farmers to help them shift their production to support plant-based products. A challenge, though, is where governments refuse to offer tax incentives to help create environmentally friendly food systems. To this end, she encourages voters to support officials who are willing to create change in food systems.

For Heather, when it comes down to it, getting folks to switch to eating more plant-based food is simple. "Be kind," she says. "Instead of judging people for what they're not doing, listen to them, be compassionate, and just offer them amazing plant-based food. And then, when their tummies are full," she says, "they'll listen to the message."

Heather's outlook on business is her outlook on life. "My whole ethos is: if there's a problem, there's a solution. Really, there's no problem. You can say there's an issue we need to look at, and we need to solve it, but problems are such a negative way to look at things," she says. "Instead, help and learn and inspire people to find a way." Heather is also averse to the idea that success is at all based on luck. "If people say you're lucky, yeah, you're lucky to be born who you were so that you can do something. But there's no luck that replaces hard work."

Chapter 10

Daniella Monet

Actress, Investor, and Cofounder of Kinder Beauty

Growing Up as a Child Star, Using Your Platform for Good, and Building Market Solutions for a More Compassionate World

> "At some point you have to start creating the change, and you have to take things to the market so that consumers can start to implement these things into their lifestyle and thrive."

By her teens, Daniella Monet had already achieved success as a TV star and singer. Yet what would have been the end goal for most was a springboard for Daniella, who has leveraged her platform to become a major player in the plant-based business world. As cofounder of Kinder Beauty, a subscription service that delivers vegan, cruelty-free, and nontoxic hair-care, cosmetics, and body-care products, an investor in the vegan food world, including restaurant chain and CPG producer Sugar Taco, and more, Daniella is living her values and helping consumers do the same.

At the root of Daniella Monet's interest in plant-based entrepreneurship and investing is a little girl with a big heart. "I was really young when I became vegetarian," says Daniella. "I was five years old and visiting a dude ranch with my family, and it really made an impact on me.

Animals were being lassoed and tied by their hands and feet, crying out for their lives, being thrown onto their backs, and being branded—and people were celebrating it. I was really confused," she recalls. "All I wanted to do was save their lives." Later that night, when Daniella's family went to the dining hall, she saw they were serving steak. "I remember them talking about how they raise their livestock," she says. After asking her dad a few questions, Daniella reached a decision: "I said, 'I'm not eating animals ever again,' and I didn't."

When Daniella was in sixth grade, years after going vegetarian, her uncle was suffering from terminal cancer and was home on bed rest. Her aunt and uncle had a chef come to the house to help prepare healthy meals for her uncle, and it turned out the chef was vegan. Because Daniella was vegetarian, they invited her to come over and learn to cook some of the meals. Daniella asked the chef why her uncle was eating a vegan diet, and the ensuing conversation focused on the nutritional benefits of plant-based eating. "The correlation between health and veganism hit me hard," she says. At the time, she was reading the book *Skinny Bitch*, which underscores the health benefits of a plant-based diet. For Daniella, going vegan seemed the next logical step, and she's been plant-based ever since. Like for many of us endeavoring to adopt the lifestyle, she says it took a long time before she felt a sense of understanding from her family, but that it did happen. "I think that there's no one right thing you can say to other people to make them see things the way you do," she says, "so you have to be prepared to really just have your own back and be confident in your decision. Ultimately, even if it's not in your immediate family, you'll probably get pushback from your friends or your coworkers." She says that while veganism is much more common now than when she was growing up, it's still important to be able to support your own journey.

Over the next few years, Daniella earned some big breaks, including a show on CBS and several on Nickelodeon. During this time, when veganism was still fairly unheard of, Daniella says she had to have uncomfortable conversations with show-production staff about not having appropriate food options for her. "Honestly, though, it was a champagne problem," she says. "I was always able to support myself." Still, she wanted to be true to who she was, and admits, "It took some time for me to settle in and have some confidence to think, 'This is how I feel, and if it's the way that I'm going to live, there's no reason to hide it.' Now, being in the business side, I own it. I'm full throttle."

For Daniella, going vegan at a time when options were extremely limited made a huge impact on her desire to become a player in the world of plant-based investing and entrepreneurship. "I hope that will be the biggest outcome of all of this," she tells me. "I want to make things more accessible, because I remember what it felt like to not have options. I remember making do with so little, and pretending I wasn't affected, and that it was no big deal, and that I'd just eat the side dishes. I'm over those days. I want options, I want them to be delicious, and I want everyone to enjoy them."

With this in mind, Daniella began viewing her celebrity as a means to a larger, more meaningful end. "Acting really gave me the tools, the confidence, and the means to invest in things that I'm passionate about," she says. "I feel really lucky that I was given a platform, because I've been able to shed light on things that I care about."

For her first step outside of acting, Daniella decided to try her hand at investing in the stock market. "Even though I was making a comfortable living, I didn't feel comfortable in my living," she says. She also wanted the opportunity to express her values through her investment dollars, which she says is a great way for anyone interested in becoming a major investor to start out.

At sixteen, knowing nearly nothing about the stock market, Daniella contacted a financial planner and started putting money into companies she knew she was interested in as a consumer and wanted to support, such as Apple and Tesla. Not long after, Daniella became an investor and founder in a variety of businesses across the vegan sector, spanning restaurants, CPG, and beauty. Because she understands the interconnectedness of animal products between the food and beauty industry, Daniella's interests have made her not only an early investor in the burgeoning space but also a visionary in the field. "Go where your dollar goes, or where you would like for your dollar to go," she says. "If you believe in it, hopefully you'll grow with the company. It's no different with investing in a company like Kinder Beauty."

Kinder Beauty is the company Daniella cofounded with fellow actor and vegan Evanna Lynch, who is best known for playing Luna Lovegood in the Harry Potter film series. "Kinder Beauty was really special," says Daniella, who encountered the idea as a result of her longtime friendship with Andrew Bernstein, with whom she'd worked when he was at PETA. After noticing the substantial market gap in vegan beauty products from both her personal experience as a mother and her professional experience as an actress, the concept made sense to her. "Andrew introduced me to Evy [Evanna Lynch], and the three of us started spinning our wheels," Daniella says. "Andrew was really determined to create something that was for profit. In his work with PETA, Andrew was constantly giving, but at some point you have to start creating the change, and you have to take things to the market so that consumers can start to implement these things into their lifestyle and thrive."

Like many founders, Daniella says she didn't really know the extent of what she was getting into with Kinder Beauty, but her attitude is that if you're going to do something, you have to do it one hundred percent: "It's been a labor of love. The startup thing is all hands on deck, and you're doing multiple jobs and doing it for little or no money for a very long time, but it's paid off in more ways than one." Daniella says that in addition to seeing market success, Kinder Beauty has been able to create an engaged community and solve a major problem in the process. "Evy and I are both in the entertainment industry, and we both had questions about the products we were using, such as which were vegan, which were cruelty free, and what was the difference," she says. Now, with Kinder Beauty, the pair have taken it to a new level, looking at how to make products clean, more sustainable, and so on. While it's a lot to take on, Daniella is optimistic: "It feels good to be in a position of growth and to have an opportunity to solve all of those problems. And with Kinder Beauty, the good thing is you'll never have to wonder what's in your products."

"The great thing about it is, it's a conversation everyone's willing to have," Daniella says. This is reflected in the market; not only are cruelty-free and vegan brands becoming big business, but even China, a market that was completely closed to these brands, is starting to open up. Echoing Daniella's insights, a 2021 report on the vegan beauty industry revealed that it is expected to be worth $20 billion by 2026. In fact, in what may be a surprise to many, the vegan beauty sector is actually growing as quickly as the vegan food sector, and could even outpace it in years to come. And, as with the future of food, there is a great responsibility for leaders like Daniella to create a more equitable beauty industry—an industry that has notoriously underserved and overlooked people of color—that is representative of founders and consumers of all backgrounds. This is a responsibility that Daniella takes seriously.

"When you're a conscious consumer and you speak with your dollars, people listen. These big companies don't want to lose your money, and we need cruelty free and vegan to be the baseline," she says. While Kinder Beauty incorporates products from large companies, they also support smaller, independent brands. "The blend of it all really makes Kinder Beauty what it is," Daniella explains. The company also gives back, contributing a portion of its profits to environmental justice, animal welfare, social justice, and other causes.

I ask Daniella where she sees gaps in the beauty market. She says that in her experience with Kinder Beauty, there are actually tons of brands and products available to them, but sometimes the barrier is simply becoming aware of them. She notes that is one of the ways their community helps. "They keep us in check, they tell us what they want, and we're all ears. When I hear there's a new brand we need to check out, I'm definitely out there doing the work," says Daniella, who runs their social media accounts herself. "There's really nothing that's totally missing, but it's about doing the work to find the proper formulations, the companies that can handle growth, and so on." So, for those newer companies in the kind and sustainable beauty sector, she says, "Don't give up, and keep doing what you're doing!" Also, she says that even though there are lots of great brands out there, there is still plenty of space in the vegan beauty sector.

Daniella says one of things that makes her work fun is working with so many women founders. Unlike areas such as CPG, which are still largely led by male founders, the emerging clean-beauty sector is dominated by women. "We speak the same language," says Daniella, "and there's just a mutual respect."

Though Kinder Beauty gets a lot of Daniella's focus, it's not her first vegan venture. That came when she became an investor in Outstanding Foods. It was tough timing because her family had just closed on a house, but, as Daniella describes it, "I just know that there are certain things that feel so right and that you don't want to miss out on, and that was one of them." She says that Outstanding Foods CEO Bill Glaser was one of the main reasons she wanted to invest, because he was willing to invest in Daniella's learning journey as a new player in the plant-based food world. "I think it's very important to find people that will bring you along for the journey and answer your questions," she says, and she found that in Bill.

Along that journey, Daniella says, she's found that she really enjoys listening to pitches. "It's inspiring hearing people's stories. I think a lot of people know what it's like to feel passionate about something and to believe in something wholeheartedly, and feel like if only they could get the right person's ear, there's no end to the possibilities. It's exciting to be the person on the other end of that."

Daniella says she hopes to leverage her success from her early investments so she can keep supporting others with their dreams. "I don't necessarily need to create these brands and these products," she says. "I love being able to support other people. I think that there are a lot of people with so much heart and soul who deserve the means and the platform to make things happen, so if I can be that person, I'm happy to do that."

I ask Daniella what appealed to her first and foremost about the companies that have pitched to her successfully. "The people," she says without hesitation. "Hands down, it was the people. When you see someone with passion, heart, and drive, and you see that they are willing to do what it takes to see this through, there's this feeling you get that you know you cannot miss this opportunity. That was the feeling I got when I met the CEO and chef of Sugar Taco, Dave Anderson." Daniella has known Dave, who was involved with Beyond Meat, since she was sixteen, so she had deep insight into his drive, humility, and work ethic. She says something I've heard echoed by many investors, which is that they weren't necessarily completely sold on the founder's idea but on the founder themselves, and on their ability to be successful. They knew this was someone, or a team, they wanted to work with. Thanks in part to Daniella's investment, Sugar Taco now operates a successful, fully plant-based restaurant chain with multiple locations across California.

Juggling roles in multiple businesses, along with a podcast, life as a mom, and on and on, sounds like something only a superwoman could handle. When I ask Daniella how she does it, she says there's really no formula. "It's almost like it's all just happening. When you care so much about certain things, you find the time and you find the energy, and you just do it," she says. "Being a mom can easily be a full-time job, but I try really hard to remember that it's just as important to support my family to survive and to thrive, and to have a sense of balance for myself. I can't lose myself in all of it. It's not easy, but I do my very best." She also says that whenever possible, she tries to marry marketing and other efforts across brands to make things a bit more seamless—and the podcast, where she talks about all of her investments, is part of that.

When it comes to recommendations for the next crop of founders, Daniella offers two pieces of advice. "Initially, if you have the means, invest in yourself—literally, with funding, and in other ways," she says. "Put yourself first. Pay yourself first. Be your biggest cheerleader. Coax yourself into believing in yourself." When you run that out and get to the point where you're about to hit diminishing returns, then it's time to look at whom you can hire to help you. If the money isn't available for that, Daniella says, then it's time to put together your pitch deck, look at who would be a good partner and whom you would like to work with, and hit the funding circuit.

Yet Daniella adds—and I agree—that it is wise to be circumspect about taking on investors unless that's truly the right move for you. There are many ways being a founder can look beyond pitching and doing funding rounds. If you can bootstrap and find ways to fund yourself, even if it means building more slowly, and perhaps passing the hat among family and friends, you could end up as a sole or co-owner of a multibillion-dollar company—like Spanx founder Sarah Blakely, who built her entire company without outside investment and recently reached a valuation of $1.2 billion. "I've always had a very minimal team," says Daniella. "For me, the fewer cooks in the kitchen, the better. That way you stay in closer touch with what's going on—and no one's going to protect yourself and your interests like you are."

As a founder, Daniella says, "You're going to have a lot of doubt, wondering if you did the right thing, if you'll see a return, if it was all worth it. But I think if you're in business for the right reasons, money will come. Find something that's solving a problem—maybe something in your life that you wish existed, or you needed more of, or isn't available at the right price point, or just isn't acceptable. Find those things and remember that you're not the only one who wants those things to exist." That way, she explains, when you start to experience doubt, you can remember there is a market, and there are other people who want those things, too.

Daniella says that in their earliest days, there were multiple times the Kinder Beauty team encountered hurdles that could have made them quit, but between the support of their community and their commitment to the big idea behind their company, they were able to forge ahead. On the other side of those hurdles, however, was a huge spike in growth. These days, it looks like things are only going up for Daniella's many brands. Still, she says, she'll stay on the lookout for the next big vegan brand and the next hardworking, full-of-heart founders she can support.

Chapter 11

Liron Nimrodi

CEO of Zero Egg

Applying a Military Approach to Reforming the Food System, Being a First-Time Founder, and Disrupting the Egg Industry

> "In my twenties, I tried to hide the fact that I'm a woman. In my thirties, I thought if I ignored it, I would be equal. Now, in my forties, I celebrate it. I say: Go with who you are. Don't try and act differently. Don't apologize. Don't underestimate. Take what you have and go and seize your opportunities."

Like many founders in the food-tech sector, Liron Nimrodi, CEO of plant-based egg company Zero Egg, is driven by a deep desire to help satisfy consumer demand for healthier, more sustainable food choices. Yet unlike most of her fellow women founders, Liron credits much of her entrepreneurial mindset and business acumen to an unlikely source: her experience serving as a base commander in the military at the age of nineteen. She says this same experience has helped her learn to navigate the complexities of operating a multinational company that's cracking the vegan egg market wide open.

In spite of some popular mythology, there isn't only one type of successful entrepreneur—they come in all shapes, sizes, and genders, and from all backgrounds. Nobody emblematizes that more than Liron Nimrodi. Liron was born in a rural part of Israel and grew up in a moshav, or communal town, which she describes as "a place where you have a lot of group culture and nature. So, I grew up surrounded by hens, sheep, and cows, in a traditional family." From there, her life took a step that is also traditional for young adults in her country, but that is decidedly uncommon for the rest of us—she joined the army.

In popular culture, military service is seldom linked to the startup world, but after building my career in Silicon Valley by working with immigrant founders in the tech industry, I was not surprised to hear Liron begin our conversation by echoing a statement I've heard many times: compulsory military service is one of the main drivers for why Israel, a nation of only nine million people, is such an entrepreneurial powerhouse on the world stage. In fact, in 2020, Israel ranked as the number one country in the world for women entrepreneurs among the fifty-eight economies evaluated by the *Mastercard Index of Women Entrepreneurs.*

By the age of nineteen, Liron was commanding a base of twelve soldiers. Military service is a requirement for high school graduates regardless of gender, and Liron says there are definite advantages to entering such an environment so young. "I think, having this stage of life in Israel, you get to experience the nature of making decisions, of being part of a team and inspiring them. It really builds your character for life. In the army, they teach you that whether it turns out to be a good decision or a bad one, you have to make one and then move forward. You also learn to think outside the box to find a way to get things done, and to take risks. You learn to understand that the nature of things is to never be sure so you have to take risks. When you're a commander and you learn at the age of eighteen what it means to explain what you need, to describe a goal, and get a team of people to follow you, it's something that builds you into a leader. After that, you're not a kid anymore."

Liron served in the Israeli army for two and a half years, after which she went on to pursue her passion and received a degree in food engineering. From school she went into product development and a role focused on making foods healthier: "I saw the processes from the inside. I saw the food that we were eating. When I saw the animal-based products and the additives, I thought, 'The food industry needs to change. What we're eating is not healthy for us now, or for future generations.' I really believe that we are what we eat and that we need to be healthy from the inside; we need to make sure we're doing that right."

Liron's next step was to go back to school to earn an MBA. From there, she began a role at Nestlé and specialized in marketing and innovation. From Nestlé she moved to SodaStream, where she served as the global product manager. These roles gave her deeper insight into consumer psychology. "I looked at the food industry from the angle of consumers' minds. I saw their behavior, and how consumers really want to change the way they eat. They want to eat healthier foods, but it's very hard to find healthy food that's tasty. That drove me to find ways to help create those foods."

Around that time, Strauss Group, a food-and-beverage company based in Israel, launched a hub for startups. Liron approached them and said that if they had a good idea for how to improve the way we eat, she was interested in building a company around it. "That was how I met the inventor of Zero Egg and started the company. That's not the typical way to start a startup," Liron acknowledges, "but in Israel, we think outside the box." Liron's journey to becoming a founder, while unorthodox at the time, is similar to the entrepreneur-in-residence concept, an increasingly common role we are seeing in the food-tech industry and beyond, where investors or accelerators develop product or company concepts in-house, then recruit founders externally to launch them.

Shortly after meeting with Strauss Group, in 2018, Liron established Zero Egg. Just three years later, they had grown to seventeen people with a product for sale to the US food-service industry: a plant-based egg available in restaurants, college and university cafeterias, and other high-volume locations. "For me, that was exactly my vision—to penetrate the US market first," says Liron.

Unlike many entrepreneurs, Liron did not gain an interest in starting her own company when growing up because she'd seen her mother, father, or grandparents do it. Her parents each had what she describes as "lifetime work," a single job they stayed at for essentially their entire careers. Though Liron opted to go another way for her own career, she says her parents were supportive and understood that "when I make a decision, I just go for it. I'm very connected to my intuition and what I feel, and if I feel that something is right, that's the way I'll act."

Though she didn't have entrepreneurship modeled for her directly, Liron says she's always been adept at managing change, a powerful attribute for a founder. "Growing up, a lot of things changed in my life," she explains. "I changed environments, I changed schools, and we moved from a small countryside to a city, so adapting to change is something in my nature. I really embrace change, and I know how to take an opportunity and leverage it."

Liron also credits her curiosity for her willingness to try new things. "I just learn from each person that I meet. I have many role models for various things I want to do. There have been many people in my life whom I've wanted to learn from, and they've enabled me to believe in myself that I could start a company." While she's happy to learn from others, though, Liron is clear that she's not trying to imitate them. "I'm taking my strength and going with it. I'm not trying to be anyone but myself, and to go with who I am. I've never been too shy to ask questions or to say that I don't know things and ask for help and guidance. I just have a target, and I know that I'll get there no matter what."

Those who work for Zero Egg are similarly laser focused, yet they're motivated by a variety of reasons. As Liron explains to me, "Everyone who joins the company wants to make an impact. There are people focused on environmental impact, there are people for whom animal welfare is an issue that burns inside them, and there are several focused on nutrition. It's all related. The way that we eat and the way that we harm animals and how it all impacts the planet are all related. Every person wants to make a difference and create a new reality and a better world."

"What I feel is that in Israel, animal welfare and the ethical aspect really drives people here to be vegan," says Liron. In my experience as a board member for various charities, I've seen that it's not uncommon for Holocaust survivors to have animal-related philanthropic organizations or to be major donors to such causes. Subsequently, it's not surprising that Israel has the most vegans per capita of any country in the world, at greater than 5 percent and growing, which is nearly twice as much as the US and the UK. In 2020, Israel was ranked as the number three most popular country in the world for vegans, just behind the UK and Australia, while Tel Aviv has been referred to as the world's vegan food capital.

With the full force of her military training, her propensity to follow her curiosity, her willingness to engage senior mentors, and the support of practically an entire country behind them, Liron and Zero Egg were perfectly positioned to lead the charge to bring great-tasting and versatile plant-based egg products to the international market. The CEO says they started out focusing on the US primarily because Israel is too small of a market for a startup. "For us, it was natural to see the US, with its large, mature market, as our first target."

I ask Liron why she had set her sights on the egg market to begin with, rather than going with another product line. "I think milk is very mature, and there is massive competition in meat," she explains. "Eggs were just left behind, and the potential in eggs is huge—it's in almost everything you eat. It's the one ingredient that vegans are still really struggling with. And it wasn't just a US issue but a global one. It was a problem without a solution, and I knew we had the ability to solve it. Being one of the few players in this field—that's a big opportunity for a startup." At the time we spoke, though dairy-replacement products such as Oatly were making a big splash, the rising demand for animal-free alternatives to eggs, butter, and similar products was actually growing much faster than these other well-developed categories. Interestingly, it also seems to be women who are among the first and fastest out of the gate when it comes to the egg category. "I think maybe it's because women see the opportunities, and they just go after it. I want to believe that's why," Liron says. "That's what it was for me."

Zero Egg now operates a US-based subsidiary with a team focused on sales, marketing, and operations, while the Israel-based parent company focuses primarily on research and development. "In the US, half of the team is remote, and we're working in different time zones. In Israel, we're starting our afternoon when the US team is waking up. To take a team and make them feel cohesive and like they want to work together is challenging," Liron confesses, "but the fact that everyone in the company believes in the goal and the vision creates the bond. That's what makes us make the effort to get over time zones and culture differences. It makes us really work as one team. Part of my goal is to make a global team working together in a culture that is open and diverse."

When I ask Liron what advice she'd offer to entrepreneurs who are just starting out, she returns to the idea of asking for mentorship and guidance: "My home base is to always find someone that does exactly what you do and just did it before, and then, instead of reinventing the wheel, you learn from others who took a similar path. This way, you can jump forward very quickly." Liron's approach is decidedly different from the fiercely independent and male-dominated founder mantra of "move fast and break things" that often echoes out of Silicon Valley—an approach that has resulted in a myriad of deeply rooted political, legal, and cultural issues that are only now coming to the forefront of our society.

Putting this mindset in action, Liron shares that when she was learning to span the culture gap between Israel and the US, Liron called on an Israeli CEO who had already become adept at communicating with Americans. "When I talk to an Israeli person, I know the body language, and I really understand their deep motivations—the Israeli nature. When you talk to someone from the US, there's a different nuance and different codes that are hard to read if you're not familiar with that. It's like getting to know a new language. We are very direct," says Liron, noting that this is likely due to how people learn to relate to one another during their military service. "Sometimes the way that we talk can sound rude to the other side, and we know that, so sometimes we can be overly polite, but then the message might not get through. That's one of the cultural differences that we needed to learn to manage."

Liron also advises paying close attention to hiring, and looking for leaders: "When we recruit, we're not recruiting according to resume or experience or brands that the person worked for. We recruit for character. We are looking for people that are hunters—the mindset is about finding solutions and looking for ways to achieve goals. People who have inner self-motivation. We look for accountability, as well. Every person in the company is the boss of their domain, so we're looking for people who are leaders." Liron says they learned from experience that when they hired based on someone's resume, choosing folks who checked the boxes, it didn't work out: "When we hired people for their nature, that was a success." Interestingly, Liron is not alone in this approach, as several notable companies, such as UPS and CVS, have dropped many of their resume requirements in the last year to stay competitive in the market.

Beyond accelerating sales and growth within the US market and fundraising, Zero Egg is looking to be the leaders in the alternative-egg market—first in the US, then globally. Once they conquer the food-service sector, they plan to launch in grocery stores. "We see a very bright future," says Liron, and one where they'll continue to penetrate deeper into the egg category. "We are all about eggs. We're not looking beyond eggs because there is so much to do in this category. We are the experts on eggs, and this is where we want to be—and where we want to be number one."

I can attest to the value of focus and perfecting a single product or product line. Frequently I've seen companies achieve some degree of success with one product, then quickly diversify and try to create something new rather than improving on what they've got. This can dilute a company's efforts—especially a newer and smaller one—as they become spread thin across multiple product lines and take on a slew of exponentially increasing new challenges all at once. "I really believe in focus in every aspect," says Liron, "being focused in business, being focused, and being really great, tasty, and offering a nutritious product in the egg category." This idea of focusing on domination in a product line or category and avoiding mission drift, known as the hedgehog concept, is a core business principle I lead with every day, and often share with founders I advise.

It's hard to imagine that, with all of this going on and being a mother of two, as well, Liron has time for family and self-care. But as she tells me, "It's possible to have a balance. My therapy is yoga." Not only does Liron practice; she is also a teacher. "That's my way of, again, getting focused. Being on my center and making sure I'm calm on the inside. That helps me, too, when I'm making a decision, to make it from my core."

Founder, CEO, military commander, yoga teacher, mom, and someone who's helping to improve life on this planet. It's hard to imagine a richer, fuller, or more meaningful life than that. Yet Liron insists she's not the exception but rather the norm—that so many women can accomplish more than they realize. "I would like to inspire other women to know that they are capable, that they are worthy," says Liron. "That's what I see among women around me—they don't estimate themselves according to their real worth. It's sad when women have imposter syndrome, because you can't fulfill your potential from this mindset."

"In my life, I've gone through stages," says Liron. "In my twenties, I tried to hide the fact that I'm a woman. In my thirties, I thought if I ignored it, I would be equal. Now, in my forties, I celebrate it. I say: Go with who you are. Don't try and act differently. Don't apologize. Don't underestimate. Take what you have and go and seize your opportunities."

Chapter 12

Miyoko Schinner

CEO of Miyoko's Creamery

Escaping from the Yakuza, Reinventing Yourself Again and Again, and Understanding the Ethics of How We Eat

> "Ultimately, it's not just about the food or the
> marketplace, or anything else; it's about what
> the responsibility is that each of us is going
> to take to make the world sustainable and
> compassionate . . . It's not going to be a technology
> that saves us; it's our own awakening."

For many entrepreneurs, starting a company is a terrifying leap—so imagine taking that leap multiple times. For Miyoko Schinner, success hasn't come easily. After achieving early success in Japan, she was forced to flee for her life and start again from scratch. A second attempt in the food industry ended with her just breaking even. Still, she couldn't shake her vision of bringing change to the food system—this time, disrupting the dairy industry. Fortunately for Miyoko and the rest of us, the third time's the charm, and today, Miyoko's Creamery, which she founded at age fifty-seven, is pioneering the plant-based cheese market, while Miyoko herself is a role model for founders everywhere.

Miyoko Schinner describes her birthplace of Hiyoshi, Japan, as a small village with just one car. "I grew up in a house with a single mom at a time when that was frowned upon," Miyoko recalls. Typically, women were married and played the role of housekeepers, she says: "They weren't strong—they were very submissive and did whatever the husband told them. But my mother was very strong, and she had her own business." Every day, Miyoko says, women would show up to the house, sit on the tatami floors, and sew stuffed animals, which Miyoko's mother would sell: "There was always fabric and stuffing and boxes everywhere." Between the women who worked as her mother's all-around helpers and those who worked exclusively with the business, Miyoko recalls, "my house was full of women." And she says it was growing up outside the boundaries imposed on most women and girls in Japan that made her a bit of a rebel.

In 1964, when Miyoko was seven years old, her mother and father—who was from the US—decided to get married, after which they moved to California, to a small three-room duplex in Mill Valley, a suburb of San Francisco. Miyoko says it was a modest lifestyle, but still a far cry from the rice fields of Hiyoshi. In California, Miyoko's parents built a stuffed-animal business together. "I guess from a very young age I had this idea in my mind that you start businesses—that's just what you do." It was rough going at first, living in an entirely White community and not speaking English. "I used to take funny lunches to school, like rice balls, which are popular now, but not then. The kids would make fun of me," Miyoko remembers. Though there was a significant Japanese immigrant population in San Francisco, Miyoko says they were isolated from it. "My mom didn't know about Japantown. Her days were filled with learning to cook American food for my dad," she says.

"Back in Japan in the 1950s and '60s, we didn't eat dairy," Miyoko recalls. "It wasn't part of the Japanese diet. Mostly we ate rice and noodles, maybe a little bit of fish or occasional chicken, but meat wasn't an everyday thing." That changed when Miyoko's family landed in the States. Her father wanted to make sure that she was "well nourished," according to American standards, and she began eating meat at every meal. "I think I fell in love with meat," Miyoko says. "It was new, and I wanted to be American." Yet the love affair was short lived—five years later, at age twelve, Miyoko went vegetarian. Though she shunned the American way of eating, it would be some time before she reconnected with her Japanese upbringing.

"I wanted to be American. From the time I set foot in the United States, for the next decade or so, I tried to forget that I was Japanese and assimilate myself," Miyoko says. "I remember going to bed wishing that I would wake up and my hair would be blonde." After college, Miyoko began to embrace her Japanese roots and returned to her home country "to try and reclaim that part of me that I felt I had lost." For a long time, Miyoko felt she might stay in Japan. "You could walk around at night and you felt safe. If you lost your purse on the subway, you'd probably get it back because someone would turn it in. You just felt like there was a safety net—that someone would look out for you."

It was in Japan that Miyoko made the switch from vegetarian to vegan. "When I went back to Japan, I lived with my aunt and uncle, initially, and my uncle was very upset that my aunt had to prepare different meals for me because I was vegetarian. In Japan, there's fish stock in everything, so she had to cook separately for me." Miyoko says she felt that she wasn't being considerate of her aunt, so she started eating fish. But again, the shift was short lived.

"Then one day I sort of had a wake-up call, and I asked myself, 'Why have I started eating fish again?'" Miyoko says that, as many do, she'd somehow stopped equating fish with other animals, and she was also eating a lot of dairy products, especially cheese. She says while dairy products weren't easy to come by, especially ones that were tasty, she did manage to find a shop where she could buy richer, European-style cheeses. Miyoko didn't have any vegan or vegetarian friends, but she did have a subscription to *Vegetarian Times* magazine: "I think what propelled me into veganism was an article about the dairy industry. I was shocked."

Miyoko says giving up her beloved European cheeses was difficult, so she started experimenting with making her own vegan versions. "I started with pureeing tofu and putting it into smoothies and making tofu ice cream. Then I started making my own soy milk, because you couldn't really buy that in Japan at the time." Then Miyoko came across a recipe for a cashew mayonnaise, which gave her the idea of making cream out of cashews. "No one was using cashews for anything in the

1980s," Miyoko recalls, but she started experimenting with them as a heavy-cream substitute for creations such as béchamel. She says when she went to the US each year, she would buy ten pounds of cashews to bring back with her. "Those cashews were gold to me because I couldn't get them in Japan at the time."

Eventually, she started hosting dinner parties every Friday night where she'd serve a twelve-course tasting menu. "I'd spend all day Thursday and Friday cooking for it. Back in the 1980s, tasting menus had just started becoming popular in Tokyo, and no one was doing them in the US yet," she recalls. Eventually, several folks from the natural-foods industry came to the dinners. One thing led to another, and Miyoko started helping cafés and juice bars design their menus, offering cooking classes at department stores, and writing articles about vegan food for magazines. She also set her sights on opening Tokyo's first haute cuisine vegan restaurant. But then things took a dramatic turn.

As Miyoko's star began its rise in the world-renowned Tokyo restaurant industry, she found a partner who was interested in opening the restaurant with her and expanding her budding empire. Unbeknownst to her, however, her partner was connected to the yakuza, the Japanese organized-crime syndicate. "Basically, he wanted to own half of me and everything I was doing. At the time I was also doing gigs in Tokyo singing jazz, and he wanted to manage my singing career, and he wanted in on half of the cookbook I was writing." Miyoko stood her ground, "and it got really ugly," she says. "People were knocking on my door at all hours of the day and night. I'd just gotten a gig teaching at one of the top cooking schools in Japan, and he called them and threatened them, and I lost that job. I really was afraid I was going to be murdered." The police told her they couldn't help her because he hadn't made good on any of the threats yet. "He said if I didn't go along with him, he would make it impossible for me ever to work in Japan again," she says.

Life became so scary that Miyoko was forced to make one of the most difficult decisions of her life: to escape for safety. "I left very carefully," she says. "I took two suitcases, and that was all." Miyoko left more than her business behind—she also left her husband, who was unwilling to leave his family in Japan and didn't want to leave his new business. Six months later, when he visited her in the US, they decided to divorce.

"I thought I was going to spend the rest of my life in Japan—I loved it. Things were really looking up with my business, and I was really looking forward to making a life there. My husband was the nicest person in the world. But I left because I had to save my life." At thirty years old, suddenly back in her adopted homeland, Miyoko had to start over.

"I decided I wanted to start a bakery because I'd had a little wholesale bakery in Japan. I started baking cakes in my kitchen, along with beautiful French-style tortes and other items," Miyoko says. She started selling them at Whole Foods and similar outlets. "I would bake three days a week; then three days a week I'd deliver from the back of my Volvo station wagon. Business was different in the late 1980s and early 1990s—you could bake something in your house and just deliver to a store," she laughs.

Miyoko says it was a struggle at first. Not having employees, she made enough to cover her bills, but that was about it. "It worked for a while," she says, "but I still wanted to open a restaurant." She began to expand her network by attending trade shows, where she met other vegans and vegetarians. She also joined the San Francisco Veg Society. Through the society and with help from friends and family, Miyoko was able to raise about $50,000—not a windfall, but enough for Miyoko to open a small bakery-café near Japantown, which she called Now and Zen. It was 1994, and that same year Miyoko married her partner, Michael; on top of being a restaurateur, she was now balancing life as a spouse and a mom.

"The restaurant was everything Anthony Bourdain wrote about in *Kitchen Confidential*," Miyoko tells me. "I had chefs who were addicted to heroin. One day, a new chef I hired walked out of the kitchen in the middle of cooking, even leaving the burners on. I'd just had a baby, and had to run down there and cook that night. I had a manager who was stealing from the till, and it turned out he was a cocaine addict. It was everything you read about."

There were bright spots, however. The cookbook Miyoko had been working on in Japan had never been published, so she set to work translating it into English. "I submitted it to the only publisher I knew who published vegan cookbooks," she says. "Two weeks later they wrote back with an offer on a contract." Her UnTurkey recipe, which I still recommend to folks today, was in that cookbook, as well as on the menu in Miyoko's restaurant: "I came up with UnTurkey in Japan in response to some friends who'd never had turkey and wanted to know what it tasted like, so I created it for them for fun."

Every Thanksgiving, Miyoko's restaurant would offer a special UnTurkey feast for carryout: "We were selling literally hundreds of turkey dinners." One Thanksgiving, a man showed up asking for a table. Miyoko told him they were all booked up. "I'll sit on a bucket in the kitchen," he told her, "I just need to eat here."

"Literally, he sat on a bucket in the kitchen and watched me cook," she laughs. "We were out of chairs." It turned out the man had a company that made a protein powder, and he was so delighted with his UnTurkey meal that he encouraged Miyoko to take her food to a trade show called the Natural Products Expo. The next year, she went to the expo, in Baltimore, Maryland. "I wrote $50,000 worth of orders in one weekend. Then I had to figure out how I was going to make and distribute them," she says. At the time, the UnTurkey's main competitor in the market was Tofurky, founded by Seth Tibbott, who, in a funny twist, would go on to become Miyoko's first investor in Miyoko's Creamery several years later.

Miyoko began faxing the orders to Cornucopia, a major food distributor, who called her and told her, "We don't know who you are—stop faxing us these orders." Miyoko agreed, then, the next day, faxed them again. Finally, the distributor did some checking and confirmed with Whole Foods and other outlets that they did, indeed, want the products, and agreed to carry them for the season. "I got a time-share kitchen and started making UnTurkeys and shipping them," Miyoko says. She decided to capitalize on the momentum by creating three more meat alternatives—UnSteakout, UnRibs, and UnChicken breasts. She rented a seventy-five-hundred-square-foot production facility, sold the assets to the restaurant, and put the proceeds into her new natural-foods company, also called Now and Zen.

Although Miyoko had experience in the food industry, it was her first foray into food distribution. "It was all new. I had to figure out, What kinds of equipment do I need to get? How do I scale?" she says.

Miyoko was also still making cakes, but her new resources allowed her the ability to ship them all the way across the country. "They were probably the first vegan cakes stores had," she recalls. Miyoko also developed a dairy-free whipped topping called Hip Whip, and got a contract to supply her vegan cinnamon rolls and chocolate-chip cookies to United Airlines.

Being first to market has its advantages, but it also has its challenges. As Miyoko describes, it was so early that there wasn't really an alt-protein market yet, and so investors didn't fully understand what she was doing or the potential. She was making $1 million per year, which may not be considered all that much by today's standards, but at the time, she says, "it was pretty sizeable revenue. But we couldn't get any capital, so I couldn't scale it. In the beginning, retailers are happy to put you on the shelf," she explains, "but then they start wanting you to spend money on promos, and I didn't have that money." Miyoko had a few helpers, but she constituted the entire sales and marketing department. (Incidentally, one of Miyoko's office managers at the time was Colleen Holland, cofounder of *VegNews*.)

There were other major challenges to deal with, as well. Miyoko's husband was busy starting his own law firm, so not only did she have her hands full with her business, but she was also raising toddlers. On top of it, her mother was dying of cancer. "I was going to have a nervous breakdown," she says. "I got an offer to buy the company. They offered me just enough to get out of debt and that was it; it was essentially an asset sale." It was 2003, and for the second time, the budding business empire Miyoko had built was no more.

After a brief stint keeping the books at her husband's law firm, Miyoko ventured into real estate, where she focused on 1031 exchanges— a tax-focused transaction that involves swapping one investment property for another. Business was good, but, having just passed age fifty, Miyoko began to feel dissatisfied: "I was at the midpoint of my life, and it felt like things were going downhill. I'd never done what I wanted to do, which was to impact the food system. I was just making money for the sake of making money, and I began to feel really empty and really depressed." In 2007, she made another switch: "I did my last exchange and I closed up shop. I didn't know what I was going to do except that I was going back to food. So, I started by focusing on cooking classes."

Over the years, she'd been playing around with vegan cheese, and she started to dive deeper.

"I decided to write a book, even though I didn't think anyone would buy it. Lo and behold, it actually sold," Miyoko says. Her book *Artisan Vegan Cheese* became a cult classic overnight, and with it, Miyoko's celebrity exploded. Still, as a woman over fifty, she doubted anyone would take her seriously. "I thought, 'There's no way I'm going to get a chance.' But next thing you know, I was traveling the country giving talks and cooking demonstrations."

Along the way, she reconnected with Seth Tibbott of Turtle Island Foods (creators of Tofurky), who encouraged Miyoko to start her own company and told her he'd be an investor. "I really didn't want to start a company because I was worried about losing people's money again, and I felt like nothing I touched could turn to gold, but he really gave me the confidence to give it a go," she recalls. Previously, she'd met plant-based entrepreneur Ryan Bethencourt, who, when he tried some of her cheese, also expressed interest in supporting a business venture, but as Miyoko says, "I just wasn't thinking that way at the time." (Check out Dr. Sandhya Sriram's story in chapter 15 to learn more about Ryan's role in the future of food.) But by the time she connected with Seth, like a fine vegan cheese, the idea of going back into business had had time to develop.

In January 2014, Miyoko started fundraising and came up with $1 million in just six weeks. Miyoko counts Twitter cofounder Evan Williams and his firm Obvious Ventures, along with Stray Dog Capital, among her early and ongoing investors, as well. (Stray Dog CEO Lisa Feria is profiled in chapter 5.) "I was kind of amazed how easy it was to raise that money during a time when people really didn't take vegan food seriously," she says.

By September of that year, Miyoko opened up shop out of a former grocery store in Fairfax, California. "We had a production facility that was only fifteen hundred square feet, but it was probably the world's biggest vegan-cheese production facility in the world at the time," she laughs. They opened with online sales on a weekend, and by Monday morning, they had $50,000 in orders. For Miyoko, it was her proof of concept that there was a real market for vegan cheese—a category that didn't really exist at the time: "Prior to that, there was no way to show that there was demand; I just knew that every time I did a demo or put something online people would go berserk, and people always said they missed cheese. I've always been on a mission to prove to the world that things are possible, and I had to prove that great vegan cheese was possible." Now, she had orders to fill. At the time it was just Miyoko and three other employees, so she scrambled to hire additional help.

Within a year, Miyoko's products were being carried by several national distributors. "Within a couple of years, it was clear that we were going to max out in our facility," she says. "Every desk had a fifty-five-gallon coconut-oil drum next to it plugged into an outlet so it would be melted, and you'd have to navigate around the oil drums to get across the room." At the time, it wasn't an option to use a co-packer—a manufacturer with extra capacity that could help them fulfill their orders: "No one had the capability to mill a nut milk, ferment the nut milk, coagulate the nut milk, age the cheese . . . We don't make processed cheese, where the technology is simple and well known. What we were doing was a whole new technology, and we weren't even sure how we could scale such a manual process. We had to change the processes several times to get it to work, so it was an innovation in continual development."

One year after the seed round, Miyoko went out for additional funding to cover her rapid expansion. Just under three years in, she had a Series B round that raised $6 million, enough for her to build a massive fifty-thousand-square-foot facility. Now, however, much of Miyoko's production, including all her butter production, is being done by a co-manufacturer. "Now we know what's needed, we've proven the concept, and we have co-packers willing to make the investment in our products," she says. "They're making the investment because they see the potential for them to grow, as well."

Among her co-packers are facilities that also process animal-based products. Some in the plant-based sector are staunchly opposed to working with such companies, and when I ask Miyoko about that, she shares an opinion with which I agree. "If we can be successful, that's more reason for these companies to invest in the infrastructure for plant-based farming and move away from animal agriculture. I also think it's naive to think we can easily find all vegan investors and strategic partners. It's radically different from where it was five or seven years ago, but we're still not at that point where those opportunities exist for every single entrepreneur. If we can take power away from animal ag, we should do that," she says. "Why not help them transition to the new economy?"

Today, Miyoko has more than two hundred employees, her products are carried in more than twenty-nine thousand outlets, and her vision is only getting bigger. Yet she knows it's a task that's far beyond what just one brand can accomplish. "We need to replace the entire animal-dairy category," asserts Miyoko. "Just like what's happened in the fluid-milk category, we need enough products that are viable on shelf, and that's going to require working with retailers and helping retailers imagine what the category could look like," Miyoko says, referring to the fact that one in three Americans regularly consume plant-based milk, which is drastically higher than any other plant-based category. "Outside of plant-based milk, even with meat alternatives, they're still not getting prime-time shelf space." Stores need to realize they need more than just one vegan cheese: "They need a vision for the future. Right now, entrepreneurs are too focused on their own technology and how they have something no one else has. There are founders who've been in the market for three to five years, and they still have only one thousand points of distribution." She's more focused on what we can get on shelves now and in the near term. "Ultimately, if we have ten years to reverse the trend of animal agriculture, we don't have five years to get on the shelf. We have to get on shelves now and take it over. So there has to be more focus on products that can get on-shelf sooner. We need technologies that are scalable now," she says. "We're racing against time to save the planet and the animals."

At the end of the day, Miyoko says, technology alone isn't going to win: "And you can't just be successful by selling your story; you have to sell the story of the whole category. You have to promote others. But there's a mindset that's focused on protecting your IP [intellectual property] and setting yourself against others in the category." She says it's going to take companies working together to get retailers to invest more heavily in plant-based categories overall. "We've only got 3 percent market penetration," she underscores. "That's it. Three percent of America has eaten vegan cheese, but more than 60 percent of households have tried a dairy-milk alternative. We need to give consumers options, and that includes supporting other companies, as well. Part of being a leader in a company isn't just about the product," Miyoko says. "We're not focusing on the bigger mission, which has to do with the vision, the marketing, the sales and the channel strategy, and helping retailers and consumers think through the problem."

In Miyoko's view, a critical component of making an immediate impact is grabbing the attention of flexitarians—people who've dialed down their focus on meat and look to include more plant-based foods in their diet. "Flexitarians aren't going to eat cheese that's made out of oil and starch," Miyoko says. "I was recently on the phone with a major European retailer that was telling me that flexitarians there have the impression that plant-based foods are healthier, and yet most of the products in the cheese category have as much saturated fats and salt and a similar lack of nutrition as dairy cheese, and so that's a problem." Miyoko says that's one of the reasons she's focused on using whole-food ingredients instead of isolates and additives to deliver nutrient-dense products that also delight the taste buds, including a new cottage cheese that includes ten grams of protein per serving.

Miyoko's leadership is laser focused—she believes it's time to step back and take a broader look at our food system, not just where it is now but where we want it to be in the future: "We have to ask ourselves: What is the food system we're trying to create? It has to be much more holistic. It's not just that technology saves the day, it's also that more people get to participate in the food system. It's the impacts that our food system here in privileged parts of the world has on parts of the world that are less privileged. It's all of that. There's so much to it." When there's more diversity in the food system, including the people who make the food, the outcomes will truly serve everyone.

"If you live in the South Side of Chicago, being a locavore means eating at McDonald's. That's what happens if the system is run entirely by privileged White people," she explains. "They think about themselves, they look at the food system from a very myopic view, and they don't think about the impact of the food system on other people." We need a food system that has opportunities for all, including smaller producers, not just large conglomerates, even if they're selling plant- and cell-based products.

"We need an awakening of consciousness," Miyoko says. "All the time, when I give talks, I explain that if people want to keep eating meat like they have been, we need more factory farms, and people are shocked. Of course, we don't want that. But when we talk about incorporating animals into a regenerative agriculture model, that's something only the privileged will have access to, and everyone else will be eating beans and rice." And we can't scale this kind of regenerative agriculture without doing more harm to the environment, because the need for extensive land for grazing would require more clear-cutting and clearing of forests and other wildlands.

To Miyoko, it all points back to the need for more plant-based products and messaging to help consumers see plant-based foods as a sustainable, equitable, and delicious solution. "Ultimately, it's not just about the food or the marketplace, or anything else; it's about what the responsibility is that each of us is going to take to make the world sustainable and compassionate. That's the only way we'll ever be able to save our planet. It's not going to be a technology that saves us; it's our own awakening."

Part of that key messaging includes the countless legal challenges associated with plant-based labeling, which Miyoko is no stranger to. In 2021, Miyoko took on the entire state of California—and won. The California Department of Food and Agriculture directed Miyoko to remove the words "butter," "dairy," and "cruelty-free" from her packaging, and to take down from her website an image of a woman hugging a cow. (Incidentally, the photo was taken at Rancho Compasión, the farmed-animal sanctuary Miyoko and her husband founded.) In response, the Animal Legal Defense Fund (ALDF), a legal-protection organization, filed a lawsuit against the State of California on Miyoko's behalf. In August 2021, Miyoko and ALDF prevailed, setting a precedent for the labeling of other plant-based products. (In the past, lobbying groups for the dairy industry have urged the US Food and Drug Administration to require plant-based-milk producers to include terms such as "imitation," "substitute," or "alternative" in conjunction with "milk.")

In 2021, Miyoko was named to *Forbes*'s 50 Over 50. When the visionary founder, who once doubted whether anyone would want to buy her products, is asked for her advice for others, she replies, "Once you stop to listen and look within, you will realize how absolutely powerful, wise, and resilient you have actually become. Don't let the self-doubt of earlier decades get in the way. You will soon realize that was all rubbish." She emphasizes that entrepreneurship isn't just an arena for the young: "Step onstage and take command; your strength will become increasingly palpable as you realize that you are now the star of your new direction in life."

Chapter 13

Sylwia Spurek, PhD

Lawyer, Former Deputy Commissioner for Human Rights in Poland, and Member of the European Parliament

Championing Women's Rights for All, Combating Hatred and Misogyny, and Using Politics for Good

"I think there's no real feminism without veganism."

Dr. Sylwia Spurek never imagined her career path would take her all the way from a small village in Poland to the European Parliament. A human-rights attorney specializing in domestic violence and women's rights, she also never envisioned that some of her greatest battles would be not over gender rights but animal welfare. Yet Sylwia is leading the charge to revamp the way European nations eat and how they treat animals, including introducing a number of controversial policies that, if implemented, could create massive change for people, animals, and the environment.

Sylwia Spurek was born in a small town in the middle of Poland. "It was a quite normal home," she says, with her parents and little sister, with whom she is very close. These modest origins, Sylwia shares, would eventually go on to help inform the broad initiatives she's undertaking today within the European Parliament, because they've given her a real understanding of the issues people face in all kinds of communities.

When she was nineteen, Sylwia left her small town for university, where she focused on criminal, administrative, international, and constitutional law. Upon graduation, she went to work for a nongovernmental organization (NGO) focused on helping women dealing with domestic and sexual violence. "I've been dealing with women's rights for more than twenty years now," she says. But, as for many of us, Sylwia's work and life were forever changed the day she made the link between women's rights and animal rights.

"I became vegan in 2015 because of my feminism," Sylwia tells me. "All of a sudden, I realized that I cannot be a human-rights defender if I ignore or accept cruelty, suffering, and injustice toward animals. I think there's no real feminism without veganism. This is about empathy. This is about justice. It's about dignity and the discrimination, exclusion, and violence that women and animals face."

Once this picture crystallized for her, Sylwia, who'd never considered a career in politics, realized that becoming a politician would give her a platform to talk about all of these issues and explain why they're related. "When I entered politics, I started to feel more independent because suddenly I could talk freely and openly about all the issues that are important to me, and that I think should be important to all of us. I couldn't speak so openly when I was a deputy commissioner for human rights or anywhere else where I was working for somebody else, representing someone else's goals and values," she says. "When I entered politics, I realized that it was an open platform for me to state all of the ideas, projects, and recommendations of NGOs, feminist activists, plus animal-rights activists, and to put all of these issues in front of actual decision makers."

Sylwia says that in her experience, activists and those working at NGOs are often ignored. "When you're a politician, all of a sudden you can be heard, and you can be very loud with your values and standards," she says, "and this is a good thing. We often think that politics is dirty, that it's not for women, that it's inefficient, and so many women don't want to try and enter. Yet in my role, now I can be a microphone for NGOs and activists."

Clearly, Sylwia has been a major advocate for women, but, as the topic often emerges in conversations with women leaders, I was curious to hear whether women have advocated for her, as well. "During my career, I've met wonderful women who've supported my goals and my dreams," she says. "My bosses in the government, in NGOs, in academia, in New York, in Europe. Lots of young women were involved in my campaign for European Parliament, and I think having young people engaged in this way was really important, because our platform was really progressive in terms of human and animal rights, and they got it. A lot of young women and men, even girls and boys, wanted to be active and support me, and I really appreciated it."

Yet there was one woman in particular who's always been there for Sylwia—her sister, Anna Spurek. "All of these years I've had a wonderful relationship with my sister," she says. "I'm so happy to have such a relationship and such a sister." In 2019, Anna took a leave from work and traveled from Italy to Poland to support her in her electoral campaign, serving as campaign chief and organizing all of Sylwia's campaign activities. "I can say that I won these elections because of her," Sylwia shares. "I know she would do anything and everything for me, and it's mutual." Now, Anna serves as the COO of Sylwia's think tank, Green Rev Institute, which works to lobby decision makers for changes in agriculture and animal farming. "We're talking about animal-free farming together," Sylwia says, "with me from my position as a member of Parliament and her from her position in an NGO."

When I ask how her work is going, Sylwia says that it's largely been an uphill battle. "I'm one of a few politicians in the European Parliament, during this term, who are vegan," she says. "We are a small group. I really think it comes down to courage. A lot of politicians from right- and left-wing parties are afraid to talk about the changes that are needed within our food systems and agricultural models. When we talk about climate change, there's no controversy in talking about energy or transport and changes and reforms we need to make there, but there's a lot of controversy when I start to talk about agriculture and our current model of food and nutrition." Still, Sylwia says, someone's got to lead the charge to start these discussions, and it's a mantle she's willing to take up. "We need someone who doesn't fear what the voters will think and whether they'll get support in their next election or not, and who doesn't worry about whether there's hate speech directed at their ideas."

Because of her willingness to be so vocal about these issues, Sylwia's experienced plenty of vitriol directed at her via social media platforms. "There's a lot of hate in my social media when I propose some of these ideas," she says. "I can say as a feminist that, being a feminist and talking about women's rights, I have not faced as much hatred as I have faced as a vegan talking about the rights of farmed animals." Sylwia says that her vegan male counterparts likely also face similar backlash, but that there are real differences in how men and women experience online trolling. In fact, she's been working on issues with gender-based cyberviolence, and she is one of the rapporteurs in the European Parliament for the legislative proposal on cyberviolence aimed at helping women feel safe online. Otherwise, Sylwia says, women can be driven out of cyberspaces, canceling their accounts and shutting down their websites out of fear for their safety.

Still, Sylwia says, when it comes to the hatred she faces online, most of it is centered on veganism and animal rights more than women's rights. "It's surprising," Sylwia says. "Being a feminist and defender of women's rights, I know it's clear some people believe the idea that women deserve equal rights is controversial. Yet talking about animal rights is even more controversial, unfortunately." When I ask about the breakdown between male and female detractors and supporters, Sylwia says that overall, women are more positive and more actively engaged in supporting her ideas. Generally speaking, men are less so when it comes to animal rights, opting for the status quo.

Sylwia says that it's not just cyber-strangers who've been critical. Many people she's worked with in human rights didn't understand her desire to include animal-rights issues in her platform. "I wish all human-rights defenders were vegans, because this is also about injustice, and we need to understand our fights are very similar," Sylwia says, but she's optimistic that in time, feminists and other human-rights advocates will begin to see the similarities with animal rights she saw.

Yet, as so many of us know, it can be frustrating waiting for this shift. "I wish change would come faster," Sylwia admits. "I dream about a world without violence and discrimination and exclusion, and I want this world here right now. It's hard to be patient. But I can see some changes," she says. "People are becoming more aware and are more interested in what they buy and what they eat." She says more people are concerned with the welfare of the animals they consume, whereas she's more interested in not eating them in the first place, "but it's a start. They're talking about bigger cages and I'm talking about empty cages, yet adopting these types of changes in treatment is an important first step."

It's a start, but Sylwia is not about to stop there. "When I entered politics and I won the election," she says, "I knew I would have five years to start some crucial debates about how we treat animals, about the injustice of exploitation and killing, and about changes that are needed in agriculture relative to environmental issues. We need to change our whole approach to these issues. Having this debate based on facts, data, and analysis is crucial right now. I don't know whether I'll be elected for the next term, or even if I would run again, but right now I have two years left to start doing what I can to make these changes in these sectors. I can be an ambassador for these changes, and that's my mission."

Sylwia is focusing on two overarching areas. The first is placing restrictions on the meat and dairy industries, directing more resources toward supporting plant-based food production, and recognizing animal rights. "We need to adopt a ban on advertising milk, eggs, and similar products to end the propaganda that's currently being presented," she says. "The consumption of meat should not be promoted, and we need to stop. At the same time, there is a huge amount of funding within the European Union directed toward these industries, and we need to redirect these dollars to support the plant-based sector, including research and development." The second area of focus for Sylwia involves removing taxes on plant-based products and adding a tax on meat and dairy products, as well as educating consumers about truly healthy eating. This would include revamping nutrition education in schools.

When it comes to animal welfare, Sylwia wants to phase out animal farming altogether, stop the use of animals in drug and other product testing, halt the exploitation of animals in the entertainment sector, such as circuses, and ban hunting and fishing. "I know it's controversial," Sylwia says, "but I am inspired by Alice Walker's view: '[Animals] were not [created] for humans any more than . . . women [were] created for men.' Animals are not objects. After all this time there are still some men who think that women were created for them, and we're still at the very beginning of the discussion around animals, so there's a long way to go. But we're starting."

In our conversation, Sylwia offered praise for plant-based founders and CEOs, acknowledging them as critical to a humane and environmentally friendly future. "I wish them determination and good health, because there's still a lot of work to be done. Don't stop, because you can have the same competitiveness as the meat and dairy industries." She also says she has a special admiration for women in these roles, "because it's still not always easy to be a woman in business, especially in the plant-based sector. So I say: Go, girls!"

When I ask Sylwia how plant-based founders and CEOs can work more effectively with politicians, she says one of the things she welcomes most is their ideas. "I want them to tell me how I can support them. I talk a lot to women who run plant-based businesses, but I would love to hear even more ideas, especially from inside industry, about how I can support their dreams, because we share similar dreams and similar values." Sylwia also says she supports any alternatives that end animal suffering, support the environment, and provide healthy, ethical food, including cell-based products. "The European Union and all such organizations should provide financial support to all startups that want to work toward these solutions. The meat and dairy sector are the businesses of the past. We need to support the businesses of the future."

In terms of the political future, Sylwia says she has a vision that someday a president of Poland will be a woman, a feminist, and a vegan. "Right now, my mission is to be a legislator and a lawyer drafting legal changes, as I was trained to do. But at the same time, right now there are no such politicians in Poland who have a chance to win. I'm not saying I have a chance, but maybe someday Polish voters will be mature enough to vote for such a candidate as president." Sylwia says she doesn't have any plans to run for president, but instead is focusing on highlighting the changes that need to be made in Europe right now.

Perhaps somewhere in Poland right now there's a girl or young woman who will someday be the president Sylwia envisions. I ask Sylwia what advice she has for the younger generation, including next-generation founders, CEOs, food scientists, and the others who will play major roles in the future of food. "Don't give up," she says. "Follow your dreams. Be the change you want to see in the world. You will find allies, more than you probably imagine. You'll get the support you need—just don't give up, because the world needs you."

Chapter 14

Priyanka Srinivas

CEO of The Live Green Co

Leaving an Indian Slum, Using Artificial Intelligence for Better Food, and Creating a Food-Tech Revolution in Latin America

> "Now my measure of success is not about the rat race
> but about the impact I'm making. Today, I feel so
> satisfied with my life. If today were my last day and
> this were my last conversation, I would be at peace."

Priyanka Srinivas never thought she'd be an entrepreneur. Yet from modest beginnings in India, where she grew up in a small home with a hole in the roof and gazed at the stars every night, she's risen to star status in the plant-based food world. Headquartered in Chile, her high-tech startup the Live Green Co is blending ancient food wisdom with modern technology to improve the health, taste, and sustainability of CPG. A first-time founder and CEO, Priyanka's story is one of bravery, perseverance, and hope.

Priyanka Srinivas was born in Hyderabad, a small city in southern India that she describes as an amazing confluence of cultures. "When it comes to food, people, practices, languages, it's all over the city." Priyanka says that when she was growing up, her mom worked, which was a rarity in her town: "She was one of ten children and said that aside from food, they didn't always have the resources they needed, so she wanted to do more, so she started working from an early age. Then she married my father."

Priyanka says her mother was also unusual in that she treated her and her older brother similarly, giving them the same opportunities. "It wasn't until later in life that I would see my friends or my cousins and realize that their education and their experiences had been cut off. At school, my mom would send me on educational excursions where I was the only girl going." Instead of spending money on their education, many families would save the money for the girls' dowries. "It seems like an ancient concept," Priyanka says, "but unfortunately it still happens in India. So, I would say a lot of my strength and my personality and the way I look at things is thanks to my mom."

"My dad is a very progressive man," Priyanka says, "but he didn't want me to be in the kitchen—my mom had to do that. My mom had to do everything—she started a business and she also had to take care of both of us. My dad was a businessman, as well," she says, "but he wasn't as good at it. So, my mom was taking care of us, making money, and paying off his debts." Priyanka says that the upshot is that her mother also made the rules of the house, which included prioritizing her children's education. "I grew up in a rented home that was less than the size of a twenty-by-twenty-foot container. We had a hole in the ceiling, and you could see the stars. My mother would say, 'I can't afford a holiday for four people, but if your school is offering an experience, I can pay for that just for you.' So, she made it possible for me to go and see the world."

When it comes to Priyanka's early relationship with food, she says their meals were always home-cooked and were always portioned. "My mom had a scooter, and she would take me along to the marketplace, where she'd buy sacks of groceries for the week. Anything processed was almost unheard of. I was probably in ninth or tenth grade before I had my first sip of Coke, and that was a luxury." Priyanka says that while her family did eat meat, that was also considered a luxury, and they might consume it only once every few months. "Food has always been whole, it has always been fresh, and it has always been homemade."

Water was also a luxury. Priyanka recalls going and collecting water and storing it in large plastic containers, then carefully measuring it out for showers and other household use. "With drinking water, you also can't get it at home, so you'd go stand in these big lines, get the water, and then my mother would boil it, cool it, and strain it."

Priyanka says that in addition to her early support for her education, her mother offered her something else important that she carries with her to this day—confidence. "Even if I didn't get good grades, my mom never made me feel bad," she says. "She just said, 'Try to do better.' She never focused on the negatives." If Priyanka had studied for twenty minutes, her mother would encourage her to imagine how much better she could do if she'd studied longer. Priyanka says that really kicked in when she was in college and she decided to apply herself fully. "Today, in my business I speak Spanish," she says, "but it's broken Spanish. Still, I speak broken Spanish with more confidence than someone who's fluent in Spanish."

Priyanka went on to university in India and completed both an undergraduate degree and a graduate degree in retail management. After graduation, her career took off, and she worked for a number of major brands both nationally in India and internationally, including Calvin Klein, FCUK, Unilever, and Target. During this time, Priyanka says she was focused almost exclusively on stepwise career growth. "For me, as a kid, success was defined as accumulation of material things, because I didn't have those things growing up. When I was in fourth grade, my sense of achievement was that I would own a sofa set one day, because we didn't have one. For me, that's privilege," Priyanka says. "My mom always told me if you get a better job, you get a better life, so my plan after college was to do really well, work really hard, and get a promotion. I did that over a long period of time, and that was my definition of success. Corporate life for me was about what I owned, where I could afford to eat, being able to travel, and so on. It feels very trivial to me today, but that's what was true for me at the time." Then everything shifted when, as Priyanka says, "the entrepreneurial bug bit me."

Similar to Suzy Amis Cameron's experience in chapter 1, Priyanka says life changed drastically with the arrival of a child—their niece. At the time, Priyanka and her husband lived in the same apartment complex as her sister-in-law and brother-in-law, and so she and her husband were part of the baby's childcare. At that point, they'd adopted a more Western style of eating, with more packaged and processed foods, Priyanka says, but when they started caring for their niece, they started paying more attention to package labels. "We started to research the ingredients, and that's when we realized that a lot of the food was pretty bad."

At the same time, Priyanka's husband was encouraging her to think about entrepreneurship. "He attended the best institutes in India, and he was the one who always wanted to do things differently. He'd already founded two startups previously and was always telling me that I was wasting my life being an employee, and that there was so much more I could do. I didn't believe him, because his definition of success was different from mine. He was more interested in purpose," she says, adding, "I was a very risk-averse person." Her husband assured her that if she had a dream, there would be people who would fund her, but to Priyanka that sounded like a fairy tale. Still, he was persistent, finally asking Priyanka to describe her dream. "I see there's a huge gap in the food system," she told him, "and I see that there's a lot of knowledge in terms of ancestral wisdom that can be applied to help deal with this."

As Priyanka describes it, she began to "ask a lot of whys." She wondered, Why are so many companies using the same unhealthy food additives? Why hasn't anyone come up with a solution yet? "Why people don't do things is usually because it's too expensive or it takes too much time," she says. "If people have to bring new novel ingredients into a formulation, it can be both time- and cost-inefficient. So, we began to wonder how we could make this process more time- and cost-effective. We knew that by using biotechnology [technology that uses living organisms or biological systems], machine learning, and artificial intelligence, you can achieve so much more." That led to the idea of taking ancestral food ingredients and applying the scientific lens of biotechnology, along with the discovery lens of artificial intelligence.

Artificial intelligence is a process that gives computers the ability to perform and learn tasks ordinarily carried out by humans. In the case of food, machine learning and artificial intelligence could be used, for example, to learn about the properties of various foods and ingredients, then use computer simulations to identify how potential combinations of said ingredients could perform in terms of taste, texture, process, and so on. Essentially, instead of trialing which ingredients to use for a new vegan cheese recipe in your kitchen, a machine could simulate and identify the recipe for you—faster, more effectively, and more deliciously. This technology is very effective, particularly in the case of the Live Green Co, when it comes to the largely unmapped plant protein and fungi kingdom.

According to the Live Green Co, there are more than four hundred fifty thousand plant species containing more than ten million natural compounds, yet less than 1 percent of these have been examined scientifically. Instead, manufacturers rely on roughly five thousand ingredients. The Live Green Co describes their recommendation algorithm, Charaka, as a "global reference point" that includes a database of natural-food alternatives that can replace synthetic, processed, unsustainable, and animal-based additives.

Historically, Priyanka says, ancestral plant wisdom is passed from generation to generation, but no one has really looked at it from a scientific angle. "We want to be a tech-first company," Priyanka says. "I don't want to build the best snack or the best burger or the best milk alternative in the world; I want to build this amazing database and the technology, show you the validation with our own products so you believe the technology works, then scale it globally with partners through licensing," she explains. "We're disrupting the food system's knowledge base." Priyanka says that machine learning will enable further discovery of new ingredients and broader applications of existing ingredients, which her company will also license.

Priyanka describes this work as helping to build more long-term, sustainable solutions to our food challenges. And she says she looks to a rather unlikely source for inspiration: the Lego Group. "Lego started as building blocks, then moved into mechanics, then into robotics. It's about the complexity that Lego can offer, and we see that opportunity in food technology," she explains. It's about taking a large systems approach and actually seeing the forest through the trees.

I ask Priyanka how they made the decision to base their company in Chile, a country she had never visited with a language she did not speak. She explains that in her role with Target, she had managed the launch of several major food brands in a variety of markets. From those experiences, Priyanka knew that she didn't have the resources to start out in a large market. Additionally, she ruled out the US because it was challenging to get to from India, and she anticipated issues getting a sufficient visa. Instead, she looked for a market where there was a heterogeneous (diverse) population that was similar to the US and had incentives that encourage entrepreneurship. Priyanka says Chile was attractive because they were offering grants and visas for foreign entrepreneurs and had free-trade agreements with more than sixty countries, among other benefits. "It also had a strong retail concentration," she says. Plus, her husband, who is her cofounder, had been in Chile several years earlier and encouraged her to consider it. Priyanka agreed, and they applied for Start-Up Chile, a government-created startup accelerator, to which they were accepted.

By that point, thanks to the international nature of her prior work, she'd become a world traveler, yet she says it was a major transition to move her household from India to Chile. There was the adjustment from the relatively conservative culture in India to the more affectionate and physically demonstrative Chilean culture. Then there was the language barrier—Priyanka landed in Chile not speaking a word of Spanish. "It was only when I got there that I realized less than 2 percent of people in Chile speak English. So I thought, 'Okay, so I have to learn Spanish!'"

Priyanka says that it's been a great choice, though, and adds that what's happening in Latin America shows that the future of food will come sooner than we may think. "Food tech is doing really well in these geographies, which is a strong indicator that even people with strong cultural backgrounds in food are making the change. In South America, people have become extremely conscious," she says. She recalls that eight or nine years ago, her husband had trouble finding meat-free options. (Once he went to McDonald's, asked for a veggie burger, and received just a bun with lettuce and tomato, with the meat patty removed.) Priyanka says that today in Chile, 10 to 15 percent of the population is now vegan or vegetarian, so options have become abundant.

One of the things that's especially exciting about this, she says, is that today's children can start with better food habits instead of having to learn them later, and that means less harm for the environment.

"People are starting to question the way food is made, either for ethical reasons, climate reasons, or health reasons. I believe that the upcoming generation will eat way better than previous generations, including mine. The next generations have become better decision makers when it comes to food."

When I spoke to Priyanka, the Live Green Co was in the midst of a fundraising round. Asked how it was going, she says while reception has been great overall, some venture capitalists are still reluctant to buy into their collaborative vision. "They want us to be a CPG brand rather than a tech company. There are a couple who have a hard time with the idea of a nontechnical CEO who's also a woman of color, running a tech company in a foreign country, but we're finding the right people to invest. Synergy of vision and ideals is extremely important for us."

Priyanka says she doesn't let doubts based on her gender get to her, in part because she's been fortunate to have so many strong women around her. When she was contemplating going out on her own, her boss at the time, also a woman, told her that she would always be welcome back and that she should go and give the company her best shot. Priyanka says this gave her the sense of support and security she needed to give the Live Green Co a go. Priyanka describes her former manager as a "superboss" who always encouraged her and nurtured her career. Priyanka recalls one incident when, after a big presentation, she pulled the speaker aside and asked a question. Her boss told Priyanka that it was a great question and that in the future she should ask questions in front of the whole group. "Every single time you have a question, raise your hand and ask it," she told Priyanka. "You don't have to overanalyze things. Never reject yourself." Priyanka says, "It was like a light-bulb moment for me. Now I'm always the first one to raise my hand."

Priyanka also credits her mother-in-law and sister-in-law. "These women have all been exceptions to the rule, and I've absorbed a lot of their values." Priyanka says that back home in India, people don't always take kindly to these attitudes and behaviors among women, however. "They can be called bullies, or are told they're being rude, but they're not. They're just doing their thing. They run households. They have control financially. My other friends work and they have a house, but they're not truly financially independent in that they don't really know what to do with their money and how to take those steps to make it work for them," she says. "These women not only make the money; they also know how to put that money to work."

When I ask Priyanka her advice for new entrepreneurs, she says, "Always be open to learning. Don't fear making mistakes. And when someone says they're not interested in investing in the company, I don't take it to heart. I remember that a lot of people must have told Google and Facebook no. I just think that those who turn me down must not be a part of my success story." Priyanka also says she's a strong believer in karma: "Karma is about doing your best and just leaving the rest, so that's what I do. Whatever happens, happens."

"Go with your calling," Priyanka adds. "Take risks and you will figure it out. No one has anything figured out on day one. Follow your dreams—you have just one life." Priyanka says that while she never envisioned becoming an entrepreneur or revolutionizing the food world, she's glad she has. "Now my measure of success is not about the rat race but about the impact I'm making. Today, I feel so satisfied with my life. If today were my last day and this were my last conversation, I would be at peace."

Chapter 15

Sandhya Sriram, PhD

CEO of Shiok Meats

**Finding Your Calling, Taking Scientific Solutions to
Market, and Creating the Future of Cell-Based Seafood**

"We all find what works for us."

*A vegetarian since birth, Dr. Sandhya Sriram has never eaten meat. And
yet in 2018, after finding herself growing tired of the slow-moving nature
of her career in academia, Sandhya decided to take a leap of faith and
jump into a little-known industry at the time: lab-grown meat. With a
focus on the dire overfishing crisis in our seas, Sandhya, along with her
cofounder, founded Shiok Meats, the world's first cell-based crustacean
company. Though they are looking to take on the world's seafood market
and today have raised $12.6 million in their Series A funding round,
getting to this point hasn't been easy, and Sandhya has faced a variety
of challenges—the biggest and earliest of which could have derailed her
career before it even began.*

"I was thrown out of a plane without a parachute": that's how Dr.
Sandhya Sriram describes the loss of her father when she was just fifteen years old. Born in India, Sandhya had spent most of her life growing up in the Middle East, but her parents moved back to India when
she was twelve. Accustomed to life in Dubai, Abu Dhabi, and Kuwait,

she found the move to be a difficult adjustment. "I loved every minute of my time in the Middle East, and I didn't want to move to India because I wasn't brought up there. It was really a culture shock," she says. Though she was raised in an Indian household, the culture in the Middle East was very different. She loved going to India for vacations, but when her family moved there, it was a difficult transition. At age twelve, Sandhya was just entering her teens. "I was the odd one out. I always stood out in groups of people; I didn't fit into the school groups. I had to change myself a bit so that I fit into the Indian student population in school, which I didn't like because I've always been a person who is very strong with my own identity, and I don't like changing for others. But I had to do that to fit in," Sandhya tells me.

Sandhya comes from the southern part of India, where, she says, people tend to put a high priority on education, and her family was no exception. "My parents were educated, and my grandparents were educated, and they, of course, expected me to be super educated," she says. "But my dad always gave me the freedom to decide what I wanted to do. He said, 'Go explore, and see what you like.'" For as long as she could remember, Sandhya says she was obsessed with the human body, learning how it works, why we fall sick, and how we can become healthy again. When at five Sandhya declared she wanted to be a medical doctor, her father bought her toy stethoscopes and other medical equipment. As an engineer himself, he'd hoped she would eventually develop a love for math, but it wasn't to be—it was biology that captured her imagination.

When they returned to India, Sandhya's father sat down with her and sketched out a map of what her educational path would look like should she choose to become a doctor. "Be prepared to study a lot of years," he told her, and together they started planning what she'd need to do academically, and what he'd need to do financially, to make it happen. Then everything changed: her father got sick.

There was no Google at the time of her father's diagnosis with kidney disease, so Sandhya had to do some hunting at the local library to find information. She also asked to accompany her father on his next doctor visit so she could ask questions. Her father's illness made her even more determined to become a doctor, and while talking with the nephrologist, she asked him how he'd chosen his field and which schools he'd attended. Yet at the same time, in the back of her mind, Sandhya began to wonder if the family would be able to afford her education.

"He was really sick for the last three years of his life," Sandhya says. The family had to spend much of the money they'd saved for Sandhya's schooling on his healthcare and, sadly, modern medicine was unable to save him. Upon her father's death, Sandhya's life turned upside down. Suddenly she and her mother, who had never had to work outside the home, were on their own. "We had to literally figure everything out," she says.

The struggle, though, made her a much stronger person and made her develop her ability to problem-solve. "India put me in situations that I never thought I'd be in. It taught me a lot about different people, about the ups and downs of life, and it taught me about how much support you can get from your neighbors, friends, and family. That's in the end what got me to the place where I am now."

Her father's death also shifted her relationship with her mother. Up until his death, Sandhya had been much closer to her father than her mother, but his passing necessitated that she and her mother find ways to lean on one another. "She needed me and I needed her—we just had to figure out a chemistry that worked between the two of us," Sandhya recalls. "One thing I admire about her now is the willpower that my mom had. She lost her husband in her early forties and had never known anything else, so she was thrown into a situation where she was drowning. She didn't want to pull me down, too, so she figured out how to swim."

As it turned out, Sandhya was right: medical school was financially out of reach—but instead of giving up, she began to search for other ways she could engage her love of biology. "I kept my mind open. Yes, I was disappointed when I couldn't do a medical degree and become a doctor. But then I had this conversation with a couple of other people who had done research and a PhD in biology, and I said, 'Why did you go for this?'" She discovered that a few, like her, had wanted to go to medical school, but either they didn't get in or it was too expensive. They told her biotech was a great option. "So I said, 'Okay, let's go for it.' We all find what works for us. And I loved it. I loved my under-grad and my master's. I loved it all. I learned a lot. I learned what not to do and what to do."

After her master's, Sandhya left India to pursue her PhD. Still, however, she wanted to make sure she wasn't too far away from her mother, because for years it had been just the two of them, and she didn't want to be out of reach if her mother needed her. "I visited Singapore, and I liked the country. I found that there was a good Indian population here—I wasn't feeling out of place. And it was closer to home," she explains.

Once a fishing village, Singapore has grown into one of the most ultramodern cities in the world. In many ways, with its fishing history and status as a major scientific hub, it was the perfect place from which to launch a high-tech company that's helping to steer the future of food toward healthier and more-sustainable options. But Sandhya still had some distance to travel in her career before she came upon that vision.

As she continued her study of biotechnology, in the back of Sandhya's mind was a persistent desire to tackle a specific disease and cure it, or make some other definitive contribution to the medical field. "But soon after my PhD I realized that not all research becomes a product or a treatment," she says. "And that's when I knew I wanted to get out eventually and set up a company, or work somewhere where we can get research into a product and into the hands of the consumer as soon as possible."

Just as the seed for wanting to pursue a science career was planted early, so was the relationship that would go on to introduce her to the idea of entrepreneurship. Sandhya met the man who is now her husband when she was just twelve years old. "We met in school and were schoolmates; we started dating when I was fifteen and got married when we were twenty-six, so we've been together for a very long time. He's been the main wall for me—someone who can absorb all my tears and my happiness and be there as a sounding board for me. He started his first business when he was twenty-one or twenty-two. When I finished my PhD and started looking for a job, he asked, 'Why do you want to work for someone? Why don't you start something on your own?'" Sandhya thought about it, but, having no business experience, she didn't know what she might do in her area. He encouraged her to go ahead and get a job, but to also start looking around for ideas as to how she might apply her knowledge and interests in her own venture. In 2014, two years after finishing her PhD and while doing a postdoctoral research fellowship, Sandhya started her first tech company—a science-based news site.

"I think most PhD graduates get to this point where they have this existential crisis and they think, 'What am I doing with my life? What's my future? Should I move into industry, should I start a company, should I stay in academia and become a professor?'" To that point, she had wanted to become a professor, but then she realized that a lot of the work she was doing was just going to become a published paper, and she wanted more. "I wanted things to happen and move at a quicker pace, and it just wasn't, and I just couldn't see myself doing the same thing for the next five years. I can't just do research, and I can't just keep publishing papers—for who? For what? That was kind of the turning point." As it happened, two of her PhD colleagues were experiencing similar feelings.

In addition to their frustration with the pace of academia, something the three scientists had in common was that when they shared their papers with family and friends, people would come back to them excited for their accomplishment but confused as to what it all meant. "So we would translate each piece of our research into simple English for them to understand," Sandhya says. The group also shared an aversion for how mainstream media tends to sensationalize scientific findings, writing something like "Cure for Cancer Found!" when in reality the study was in mice, and it would take at least another decade to see if there was any benefit to humans. "That's not portraying news in the right way," Sandhya tells me. "You're giving hope to people who are sick, but the cure is ten or twenty years away. Then an idea struck. We were like, 'Why not start a blog and start writing science in simple English and demystify biotech for people?'"

Sandhya had always been a strong student, and though she has a photographic memory, she says the way she studied was not just to memorize information and spit it back out on the test. "I've always been an overachiever and super ambitious. My style has always been to understand every sentence, every concept, and then write it in a way that the examiner can understand." This skill set helped her greatly when it came to the blog. "We started desensationalizing and demystifying science news with the truth," she says. "As part of that, we started interviewing startups and accelerators and incubators, and it was interesting." After just three months, with roughly ten thousand readers already, one company approached them and asked if they could place an ad on the blog. That was when the team realized they could monetize their creation. "We had no clue about fundraising. We had no clue about equity and shares." At that point, Google *had* been created, so they turned to the web for answers, and Biotech in Asia was born.

This initial entrepreneurial foray was a critical learning experience for Sandhya. "I learned everything that can go wrong with a startup, and everything that can go wrong with fundraising and scaling up and being an entrepreneur and all of that. Later, when I started Shiok, I definitely didn't know what I needed to do, but I definitely knew what not to do."

It was also through Biotech in Asia that Sandhya was first exposed to the idea of cell-based meat, when Dr. Mark Post showcased the first ever cell-based burger. Having studied stem-cell biology all through her schooling, she immediately became interested. As she tells me, "My mind was super obsessed with the concept. It was food that people would eat and enjoy and could become a consumer product in just five to ten years. I've been a vegetarian by choice throughout my life, and I always questioned why people want meat and seafood. I still don't know the exact answer, but I'm figuring it out along the way. But at the time I thought, 'This sounds great. Why not use stem cells?' Because if you take a piece of meat, it's made up of cells and stem cells and tissue, blood, connective tissue, fat, muscle, and all of it. What this guy's doing is taking those cells and growing it out. At the time I was trying to do that to treat diabetes and cancer and other diseases."

Her curiosity growing, through Biotech in Asia, Sandhya was able to interview representatives from companies including Memphis Meats (now Upside Foods), Modern Meadow, Clara Foods (now the Every Company), and others. That's when she came across IndieBio, the Bay Area bio-accelerator, and met scientist, entrepreneur, and IndieBio cofounder Ryan Bethencourt. Bethencourt is also a cofounder and CEO of vegan dog-food company Wild Earth, a partner at Babel Ventures, and the former head of Life Sciences at the Xprize foundation. "Shiok started kind of flowing from there," she says.

Sandhya knew that in the next three to five years, she wanted to start a biotech company that centered on stem cells. At the time, she wasn't thinking at all about disrupting the meat or seafood markets, but, again, she kept her mind open. She knew she wasn't enjoying practicing bench science anymore and wanted to be an entrepreneur. "But for me to be an effective head of a company, I needed to learn the business side of things," she says. "So, I quit being a scientist in 2015 and took up a business-manager role at a research institute in Singapore. I practically did an MBA on the job—I learned finance, budgeting, IT, patents, and communications on the job for about two and a half years."

When Sandhya met Ryan, she says they "hit it off from day one. He ended up being this informal mentor and advisor, and we started ideating a lot on what we could do together. This was in 2016, and he was already a supervisor for Biotech in Asia. I decided to quit, and I reached out to Ryan and asked if he wanted to do something more together." They decided to start a program like IndieBio in India, and Sandhya undertook a yearlong fact-finding mission. "Unfortunately, we didn't set it up, because we couldn't raise the funds. People couldn't understand the concept of IndieBio in India or in Asia, because here in this part of the world we don't have the biohacker mentality."

As a sidenote, it turns out that Sandhya and Ryan were just early on their idea. Fortunately, folks have since caught on in Asia, with over $230 million invested in alternative protein in 2020 alone. Sandhya agrees that not only was their timing premature, but also people at the time didn't know who she was—she hadn't yet built a reputation in the biotech community: "I didn't have a long career behind me, so people were questioning who I was, coming in and asking them for $5 million to set up an accelerator in an area where there have traditionally been no accelerators." They believed a US model would never work in Asia, and that companies would be reluctant to join it.

Sandhya adds, "I have always felt my age has been an issue more than my gender. I've always felt I've had to overprove myself and show that even though I'm so young, I have achieved so much in my life— look at my background." Some also questioned why she wanted to run the business rather than focus on the science in her career as an entrepreneur and were skeptical that she hadn't earned a formal MBA. "I told them I did an MBA on the job, and to ask me anything about finance, IT . . . and that if I didn't know I'd tell them, and I could go and learn, but that I could read legal documents and spreadsheets and balance and cash-flow sheets, and I could draft them and get the numbers right. But it took a while to convince people that I'm capable of doing those things."

I ask Sandhya if she was ever daunted by the fact that so many of the biotech founders she'd seen and spoken with had been men. Her response surprises me. "It never crossed my mind," she says. "For me, personally, I have never been put down as a female scientist. I should say that Singapore and Asia in general are a bit more open to women in science. Interestingly, I get this question a lot from North America. I always say that it feels great to be an entrepreneur and a scientist, but I don't know about the female part. I have a sense that it shouldn't matter, but I understand the question. Up until now it hasn't mattered, and I hope it doesn't in the future."

In my own experience talking to women founders, some have said they've been challenged on their gender, with funders telling them they'll never accomplish what they're setting out to because they are women. Some have been told they're too young, while others have been told they're too old. (See my interview with Miyoko Schinner, chapter 12, who did her first funding pitch when she was fifty-seven.) I call this the Goldilocks phenomenon.

After their failed attempt, in June 2018, Ryan began to encourage Sandhya to consider leaning on her background as a cell biologist. When she explained she was no longer interested in doing stem-cell biology, he said she could focus on being the CEO and setting up the company and bring on someone else to focus on the science. He asked her if she would be more interested in cell-based meat or in organogenesis, or making organs outside of the human body for transplant. "I said that I'd found cell-based meat really interesting, and he said, 'There you go. Take a week to think about it and let me know what you want to do.' Within the next week I had ideated Shiok Meats—everything was set in my head." She did a search to find out the most-consumed protein in Asia, where roughly 60 percent of the world's population lives. When the results showed shrimp, and when further searching revealed that no one else seemed to be working on cell-based shrimp, Sandhya had her answer. Even though she'd never eaten shrimp in her life, she would create a company that made sustainable, healthy shrimp for people who do.

"I had no idea if there was research or if there were stem cells out there that were available. I called Ryan and said, 'Ryan, I'm going to do shrimp.' He said, 'I was going to suggest the same thing!' I said, 'Awesome, do you want to invest?' That was my first pitch." When he told Sandhya how much he was willing to put up as seed funding, she told him to double it and he'd be her first investor. "Within four weeks we'd established the company, I'd found a cofounder, and we had his money in, and the rest is history."

Wild shrimp are typically caught by trawling, a method that involves pulling a massive net through the water and results in a staggering amount of bycatch. It's estimated that for every pound of shrimp caught by trawlers, between five and twenty pounds of sharks, rays, starfish, turtles, and other animals are caught, as well. In addition to environmental impacts, there are human health challenges associated with eating shrimp and other sea life, including the proliferation of antibiotic-resistant bacteria. Despite this, in 2019, the global shrimp market reportedly took in more than 4,200 metric tons, with the US alone estimated to import 110 million pounds of shrimp each year—a market Sandhya is hoping to disrupt.

When it comes to acquiring information and skills necessary to be a successful entrepreneur, Sandhya says that for her, learning on the job has been more effective than formal schooling. "I've always felt that. I contemplated doing an MBA, and then I looked at the coursework and felt that it was too theoretical for me, and I wouldn't fit in. I thought that if I started, I would leave the course within three to six months." It was the experience and learning journey of starting two companies that taught her everything that could go wrong, along with everything that could go right.

Sandhya says that one of the mistakes she made when starting her first company, Biotech in Asia, was cofounding it with her best friend at the time. The two lost their friendship over the venture. "We both were very different in work styles," says Sandhya, "and we didn't gel at all." That was the reason Sandhya decided to leave, stepping away for the good of the company and leaving her friend in charge. When it came time to find a cofounder for Shiok, she was more strategic. "I was very scared about finding a great cofounder. I knew the company needed a super strong founding team and someone who could get through a lot. We were the first cell-based meat company in Singapore in Southeast Asia. I needed a cofounder who was super understanding, and someone who could balance me out."

As someone who is outspoken about her ideas, Sandhya needed someone who felt comfortable encouraging her to step back and let them own their space. When she ideated Shiok, she put out a call through several WhatsApp groups, and one of the people who responded was Ka Yi Ling, an accomplished scientist who loved biology. "Within a week we kind of just hit it off. I found that she was open to listening, and I was open to listening to her." Not only did Ka Yi love being in the lab, but she was also open to learning the business side. "I think what works is that I understand whatever she talks about in science—the exact tech," Sandhya says. "I think that really helps on the CEO side. I can say I want a cell to grow in two days, but as a scientist I can understand why it might need four, so then I can say, 'Okay, go ahead and take four days.' It's easier on the team."

When I ask her if she has any advice for women founders, Sandhya encourages, "Just go for it. Try it and just do what you want to do and see if you like it, first of all, and if you really want to be an entrepreneur. The second thing is to make sure you have a strong support system around you, especially if you have a family and a child." Sandhya says it's essential to have an open conversation with your partner, where you tell them they'll see less of you in the next year or so and ask if that's okay with them, and also ask if they'll be willing to contribute extra to the family if you have a child. When she started Shiok, her son was five years old, and she had to explain to him that she'd be very busy the next few years setting up her business. "I told him, 'Papa is going to take care of you more.' He kept asking why, and I explained that I wanted to help people eat healthy, and he said, 'Okay!'" Reflecting on that conversation, Sandhya continued, "Women, especially, think a lot, but we hardly speak. We need to communicate."

In 2021, Shiok focused on scaling up the tech to lower the cost of their products, shoring up the company culture, values, and mission, focusing on consumer perception and education, and regulatory approval. She says it helps considerably that Asia is more open to alternative proteins and cell-based meats than other parts of the world. "Consumers there are more understanding and want to know more, and I think that's the most important thing, that we as cell-based meat companies have to be transparent with our tech and process and products, and then consumers will learn a lot more and want to adopt it." According to focus surveys, more than 70 percent of respondents want to try cell-based meat, driven primarily from the environmental and sustainability point of view, especially among Gen Z and millennials.

Additionally, Sandhya says she's seen a huge increase in investment in cell-based proteins since the start of the pandemic, with investors and major food supplies coming on board in a big way. Shiok was startup accelerator Y Combinator's first cell-based-meat investment.

"The alarming thing is that there's not going to be any more seafood left in the world in the next decade in our oceans and in our farms. So, if people still want to get those products they love, the only option is cell-based seafood. The world is soon going to have ten billion people in it, and for us to be healthy and have access to food, there has to be a mix of cell-based and plant-based protein out there." And thanks to Sandhya's vision and persistence, we will have it.

Acknowledgments

My life has taken many twists and turns over the years. I'd like to acknowledge the many people who have played a part and have helped me reach this point.

- First off, thank you to everyone who helped make this book a reality, particularly each of the incredible women who shared their innermost dreams and hopes with me for each chapter.
- To Kelly Madrone, I am deeply grateful to you for helping me bring my vision to life and for believing this story needed to be told.
- To Stephanie Lind and ESA, without your support and championing of women, I would not have been able to make this book.
- To Miyoko Schinner, you are a woman of endless inspiration and someone I am ever so fortunate to call my friend. Thank you for believing in me and VWS.
- To the PopSockets team, thank you for your support and your courage to lead businesses toward a kinder world.
- To everyone who has supported VWS and me over the years, I appreciate you and I am forever grateful for your encouragement and motivation.
- Though many have not been so lucky, I was surrounded by powerful women from the very beginning of my career. Thank you to Deborah Bowie and Karen Slevin, two of my earliest champions, who mentored me, cared for me, and helped me become the leader I am today.

- Lastly, to my mum, Maya; my dad, John; my sister, Jes; my husband, Pavle; my two pups, Vexy and Alfie; and the rest of my family, thank you for never trying to fit me into a box, for always supporting my dreams, and for loving me unconditionally.

Sources

Introduction

Jan Whitaker, "Early Vegetarian Restaurants," *Restaurant-ing Through History*, March 15, 2015, https://restaurant-ingthroughhistory .com/2015/03/15/early-vegetarian-restaurants-2/.

Tamara Omazic, "What Women Want: Female Consumers Offer Enormous Purchasing Power, Making Them a Critical Piece to the Future of Quick Service," *QSR*, March 2014, https://www.qsrmagazine.com/consumer -trends/what-women-want#:~:text=Women%20in%20the%20U.S.%20 today,the%20breadwinner%20in%20her%20family.

"Got Milk? African Americans and Lactose Intolerance," BlackDoctor.org, accessed November 30, 2021, https://blackdoctor.org/african-americans -lactose-intolerance/.

Falon Fatemi, "The Value of Investing in Female Founders," *Forbes*, March 29, 2019, https://www.forbes.com/sites/falonfatemi/2019/03/29 /the-value-of-investing-in-female-founders/?sh=15634da65ee4.

Chapter 1: Suzy Amis Cameron

"James and Suzy Cameron Invest in Plant-Based Business," *Vegconomist*, June 8, 2019, https://vegconomist.com/marketing-and-media/james -and-suzy-amis-cameron-invest-in-plant-based-business/.

Shannon VanRaes, "New Zealand Grass Greener, but Milk Prices Are Sour," *Manitoba Co-operator*, November 12, 2015, https://www.manitoba cooperator.ca/news-opinion/news/new-zealand-grass-greener-but-milk -prices-are-sour/.

Rick Barrett, "'You Have This Burden That You Carry': For Dairy Farmers Struggling to Hold On, Depression Can Take Hold," *Milwaukee Journal Sentinel*, December 31, 2019, https://www.jsonline.com/in-depth/news /special-reports/dairy-crisis/2019/12/31/

farmers-suicide-sheds-light-depths-stress-during-dairy-crisis
/2738530001/.

"What's the Average Life Span of a Cow?" *Neeness*, accessed November 29,
2021, https://neeness.com/whats-the-average-lifespan-of-a-cow/.

Chapter 3: Lisa Dyson, PhD

Kim Ann Zimmermann, "Hurricane Katrina: Facts, Damage and Aftermath,"
Live Science, August 27, 2015, https://www.livescience.com/22522
-hurricane-katrina-facts.html.

Helen Adams, "Gen Z's Changing Food and Drink Habits," *Food Digital*,
July 27, 2021, https://fooddigital.com/food/gen-zs-changing-food-and
-drink-habits.

Alec Tyson, Brian Kennedy, and Cary Funk, "Gen Z, Millennials Stand Out
for Climate Change Activism, Social Media Engagement with Issue,"
Pew Research Center, May 26, 2021, https://www.pewresearch.org
/science/2021/05/26/gen-z-millennials-stand-out-for-climate-change
-activism-social-media-engagement-with-issue/.

Kate Yoder, "Good News: The Media Is Getting the Facts Right on Climate
Change," *Grist*, August 20, 2021, https://grist.org/science/good-news
-the-media-the-facts-on-climate-change-bothsidesism/.

"Lisa Dyson's Mission to Make Air-Based Meat—and Why It Matters So
Much," *Food Technology Magazine*, September 1, 2021, https://www
.ift.org/news-and-publications/food-technology-magazine/issues/2021
/september/features/lisa-dysons-air-based-meat.

Chapter 4: Michelle Egger

Julia Wuench, "Got Milk! Biomilq Is the First Company to Create Human
Breast Milk in a Lab," *Forbes*, June 1, 2021, https://www.forbes.com
/sites/juliawuench/2021/06/01/got-milk-biomilq-is-the-first-company
-to-create-human-breast-milk-in-a-lab/?sh=7c4a48c943cd.

Beandrea July, "From Breastfeeding to Beyoncé, 'Skimmed' Tells a New Story
About Black Motherhood," *NPR*, February 11, 2020, https://www.npr
.org/sections/health-shots/2020/02/11/801343800/from-breastfeeding
-to-beyonc-skimmed-tells-a-new-story-about-black-motherhood.

Chapter 5: Lisa Feria

"Immigrant Entrepreneurs," *The Immigrant Learning Center*, accessed November 29, 2021, https://www.ilctr.org/promoting-immigrants /immigrant-entrepreneurship/.

"52% of Startups in Silicon Valley Founded by At Least One Immigrant," *Becker & Lee LLP*, accessed November 29, 2021, https://www.blimmi gration.com/52-startups-silicon-valley-founded-least-one-immigrant/.

"The State of Black and Latinx Digital Founders," *Digitalundivided*, accessed November 29, 2021, https://www.projectdiane.com.

"Cultured Meat and Future Food Podcast Episode 04: Lisa Feria," *Cultured Meat Future Food*, May 2, 2018, https://www.youtube.com/watch?v =1vE5fFTVG7Q.

Chapter 6: Shama Sukul Lee

"Sunfed CEO Shama Sukul Lee's Keynote at the 2020 City of Sydney VEP Event," *Sunfed*, August 8, 2020, https://www.youtube.com/watch?v =cGjiWAs6ZKQ.

Hemi Kim, "Factory Farming: What the Industry Doesn't Want You to Know," *Sentient Media*, August 4, 2021, https://sentientmedia.org /factory-farming/.

Victoria Baker, "Meet Shama Sukul Lee: The Sunfed CEO Making Chicken-Free Chicken," *Vogue Australia*, April 15, 2019, https://www.vogue.com .au/vogue-codes/news/meet-shama-sukul-lee-the-sunfed-ceo-making -chickenfree-chicken/news-story/4c2e31e593fb663e00a70557d437a59e.

"A New Generation of Meat," *Sunfed*, accessed November 29, 2021, https://sunfed.world.

Chapter 7: Fengru Lin

"The Singapore Water Story," *sgPUB*, October 6, 2014, https://www.youtube .com/watch?v=5BGUT7BjPl0.

Sue King, "Data Say . . . Dairy Has Changed," *US Department of Agriculture*, July 29, 2021, https://www.usda.gov/media/blog/2020/06/18/data-saydairy -has-changed.

Brandon Keim, "Farmed Chickens Can't Walk; Just Grow Them in Vats
    Already," *Wired*, February 6, 2008, https://www.wired.com/2008/02
    /chickens-cant-w/.
"Fengru Lin: The Sustainability and Alchemy of Lab-Grown Milk," *The
    Wealth Alchemist*, accessed November 29, 2021, https://the-wealth
    -alchemist.com/fengru-lin-the-sustainability-alchemy-of-lab
    -grown-milk/.

Chapter 8: Riana Lynn

"Alumni Spotlight: Riana Lynn '08," *The University of North Carolina at
    Chapel Hill*, accessed November 29, 2021, https://acred.unc.edu/alumni
    -spotlight-riana-lynn/.
"Food Scientist and Technologist Riana Lynn Breaks Down Food Tech,"
    Karen Hunter Show, January 15, 2019, https://www.youtube.com
    /watch?v=c048Q2dqKFo.
"CEO Riana Lynn on Journey Foods and the Future of Food," *Karen Hunte Show*,
    October 13, 2020, https://www.youtube.com/watch?v=7Lv-NmhK3pI.

Chapter 9: Heather Mills

Maria Chiorando, "Number of Vegans in Britain Skyrocketed by 40% in
    2020, Claims Survey," *Plant Based News*, January 8, 2021, https://plant
    basednews.org/culture/ethics/vegans-in-britain-skyrocketed/.
Katie Jahns, "The Environment Is Gen Z's No. 1 Concern—and Some
    Companies Are Taking Advantage of That," *CNBC*, August 11, 2021,
    https://www.cnbc.com/2021/08/10/the-environment-is-gen-zs-no-1
    -concern-but-beware-of-greenwashing.html.

Chapter 10: Daniella Monet

"Vegan Food Market Size, Share and Trends Analysis Report by Product
    (Dairy Alternative, Meat Substitute), by Distribution Channel (Online
    Offline), by Region (APAC, CSA, MEA, Europe, North America),
    and Segment Forecasts, 2019–2025," *Grand View Research*, accessed
    November 29, 2021, https://www.grandviewresearch.com/
    industry-analysis/vegan-food-market.

Jemima McEvoy, "Oprah Winfrey, Reese Witherspoon Back Shapewear Firm Spanx at $1.2 Billion," *Forbes*, November 18, 2021, https://www .forbes.com/sites/jemimamcevoy/2021/11/18/oprah-winfrey-reese -witherspoon-back-shapewear-firm-spanx-at-12-billion-valuation /?sh=30856294236d.

Chapter 11: Liron Nimrodi

Karen Gilchrist, "Israel and the US Among the Best Places to Be a Woman Entrepreneur," *CNBC*, November 23, 2020, https://www.cnbc.com /2020/11/23/the-best-places-to-be-a-woman-entrepreneur-include -israel-and-the-us.html.

Abigail Klein Leichman, "Israel Has Most Vegans Per Capita and the Trend Is Growing," *Israel21c*, March 26, 2017, https://www.israel21c.org /israel-has-most-vegans-per-capita-and-the-trend-is-growing/.

Zachary Keyser, "Israel Ranks 3rd on List of 2020's Most Popular Countries for Vegans," *The Jerusalem Post*, September 8, 2020, https://www .jpost.com/israel-news/israel-ranks-3rd-on-list-of-2020s-most-popular -countries-for-vegans-641412.

"Tel Aviv—World's Vegan Food Capital," *Israel*, December 29, 2015, https:// www.youtube.com/watch?v=aWxfOGtACGY.

Lauren Weber and Chip Cutter, "Help Really Wanted: No Degree, Work Experience or Background Checks," *The Wall Street Journal*, November 6, 2021, https://www.wsj.com/articles/help-really-wanted-no-degree -work-experience-or-background-checks-11636196307.

Chapter 12: Miyoko Schinner

Anna Starostinetskaya, "Miyoko's Creamery Wins Lawsuit Against California, Setting Precedent for Using 'Dairy' Labels on Vegan Food," *VegNews*, August 11, 2021, https://vegnews.com/2021/8/california -court-sides-with-miyoko-s-allowing-free-speech-lawsuit-against-the -state-s-food-and-agriculture-department-to-go-forward.

Anna Starostinetskaya, "Desperate Dairy Pushes FDA to Require 'Imitation' Labels on Vegan Milk," *VegNews*, February 22, 2019, https://vegnews .com/2019/2/desperate-dairy-pushes-fda-to-require-imitation-labels -on-vegan-milk.

Elana Lyn Gross, "The Forbes 50 Over 50 Share Their Best Advice for
Supercharging Your Career After 50," *Forbes*, June 2, 2021, https://www
.forbes.com/sites/elanagross/2021/06/02/the-forbes-50-over-50-share
-their-best-advice-for-supercharging-your-career-after-50/?sh
=1ac033c341a3.

Chapter 14: Priyanka Srinivas

"What is Charaka?" *The Live Green Co*, accessed November 29, 2021, https://
www.thelivegreenco.com/charaka.

Chapter 15: Sandhya Sriram, PhD

"Asia Population 2021," *World Population Review*, accessed November 30,
2021, https://worldpopulationreview.com/continents/asia-population.
Sonalie Figueiras, "APAC Alt Protein Startups Raised More Than US$230
Million in Past 12 Months," *Green Queen*, November 18, 2020, https://
www.greenqueen.com.hk/exclusive-apac-alt-protein-startups-raised
-more-than-usd-230-million-in-past-12-months/.
Katherine Martinko, "Why It's a Good Idea to Stop Eating Shrimp,"
Treehugger, May 5, 2020, https://www.treehugger.com/shrimp-may-be
-small-their-environmental-impact-devastating-4858308.
Michael Pellman Rowland, "Shiok Meats Raises $4.6 Million Seed Round
to Develop Cell-Based Shrimp," *Forbes*, April 28, 2019, https://www
.forbes.com/sites/michaelpellmanrowland/2019/04/28/shiokmeats
/?sh=576272e11f7c.

Index

About the Author

© 2022 Justin Chee

Jennifer Stojkovic is a multitalented executive leader in food technology and investment, policy, and advocacy. She is the founder of Vegan Women Summit (VWS), a global platform of more than forty thousand women founders, investors, and advocates dedicated to building a kinder, more sustainable world. In addition to this, Jennifer built her career as a leading advocate for tech policy and innovation in Silicon Valley, working with the world's tech giants, including Google, Microsoft, Uber, and others. She is a frequent contributor to *Rolling Stone*, an internationally featured public speaker, and she regularly appears in national press, including the *Washington Post, Bloomberg*, and *Politico*. A Canadian by birth, Jennifer is an avid athlete, a rescue diver, and an ethical vegan. To learn more, go to: jenniferstojkovic.com.

Printed in Great Britain
by Amazon

84047965R00099